MODERN
MANAGEMENT
of the High-Technology
Enterprise

PRENTICE HALL INTERNATIONAL SERIES
IN INDUSTRIAL AND SYSTEMS ENGINEERING

W. J. Fabrycky and J. H. Mize, Editors

JOHN E. GIBSON

Commonwealth Professor
University of Virginia
Charlottesville, VA

MODERN
MANAGEMENT
of the High-Technology
Enterprise

PRENTICE HALL
Englewood Cliffs, New Jersey 07632

Library of Congress Cataloging-in-Publication Data

Gibson, John E.
    Modern management of the high-technology enterprise / John E.
Gibson.
        p.   cm. -- (Prentice-Hall international series in industrial
and systems engineering)
    Includes bibliographies and index.
    ISBN 0-13-596156-4
    1. High technology industries--Management.   I. Title.
II. Series.
HD62.37.G53 1990
620'.0068--dc20   89-33436
                                                            CIP

Editorial/production supervision and
  interior design: Gretchen K. Chenenko
Cover design: Wanda Lubelska Design
Manufacturing buyer: Mary Noonan

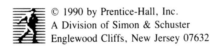 © 1990 by Prentice-Hall, Inc.
A Division of Simon & Schuster
Englewood Cliffs, New Jersey 07632

Printed in the United States of America

10 9 8 7 6 5 4 3 2 1

ISBN   0-13-596156-4

PRENTICE-HALL INTERNATIONAL (UK) LIMITED, *London*
PRENTICE-HALL OF AUSTRALIA PTY. LIMITED, *Sydney*
PRENTICE-HALL CANADA INC., *Toronto*
PRENTICE-HALL HISPANOAMERICANA, S.A., *Mexico*
PRENTICE-HALL OF INDIA PRIVATE LIMITED, *New Delhi*
PRENTICE-HALL OF JAPAN, INC., *Tokyo*
SIMON & SCHUSTER ASIA PTE. LTD., *Singapore*
EDITORA PRENTICE-HALL DO BRASIL, LTDA., *Rio de Janeiro*

To Nancy

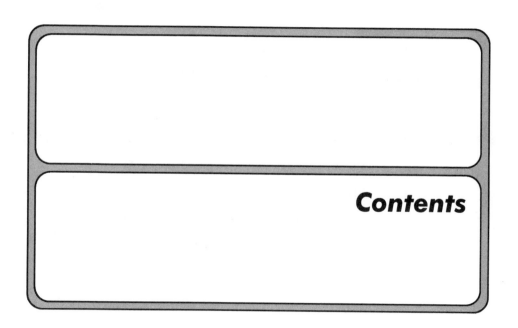

Contents

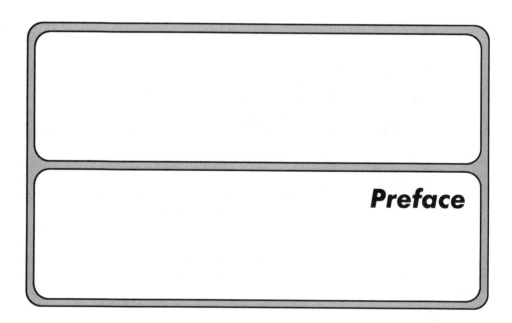

Preface

This text is a one-volume introduction to the elements of modern management. I hope to encourage professionals without specialized business training to understand the principles used by professionally trained managers to guide the typical American industrial enterprise. Because most professionals will be employed in the private sector, it is imperative that they understand the rules by which organizations, and therefore their careers, are governed.

The material divides itself naturally into four major parts: first, a discussion of the strategic environment faced by the modern American industrial enterprise in developing and marketing a product or service; second, how managers exert financial control of the enterprise; third, the management of manufacturing operations; and finally, a section on entrepreneurship. Two appendixes complete the book. In the first of these parts I begin with a short historical introduction to the American business enterprise and then move in Chapter 2 to the current U.S. strategic business environment and its relation to world markets. In Chapter 3 I consider new product definition and marketing from the same general management viewpoint. Chapter 4, Venture Analysis, takes a more analytic view of the same subject.

Whereas Part I focuses on strategy, Part II is concerned with tactics. The budget is a short-term business plan, and short-term results are indicated in the annual report. In Chapter 5 I introduce the basic language of the business world through elementary analysis of the financial statements in a typical annual report. I create a mythical industrial organization, Ajax Metal Products, Inc., and compare its financial results with those of actual industrial concerns. Each of the normal financial entries is defined, and values of the standard ratios are calculated for a recent year.

The focus of Chapter 6 is managerial economics and relevant cost analysis as the basis for operational decision making. Chapter 7 considers long-term economic decisions in which the time value of money must be taken into consideration.

Part III consists of several chapters on the management of manufacturing operations. Chapter 8 begins with an introduction to inventory control, especially modern Japanese refinements of good American practice. Chapter 9 is a brief introduction to production control, including quality control and the automatic factory. Chapter 10 introduces personnel management and quality of work life. Theories X and Y, as well as the Japanese approach to quality circles, are covered, and the Scanlon plan for participative management is introduced. The Hersey-Blanchard four-mode theory is emphasized.

Part IV consists of Chapter 11 on entrepreneurship and the small business enterprise, which may be viewed as the capstone of the text because the small, entrepreneuring manufacturer faces all the challenges of an established business in addition to several others. In a small business, cash flow is often paramount. Limitations on available cash in a rapidly expanding organization will often prevent otherwise impeccable decisions from being implemented.

I have made every attempt in this text to offer appropriate recent experiences of real business organizations as examples, because they add a realistic flavor to the material. However, I hope the reader won't misinterpret the point of "almost contemporaneous" material. One shouldn't get caught up in the flow of the narrative and forget the real purpose of the discussion. Certain recent past crises faced by Apple Computer and Harley-Davidson, to cite two of many examples, contain lessons of permanent value. Furthermore, the current condition of these and other firms used in examples may not be relevant to discussion of past crises. I have picked these companies in crisis and data for that period because they illustrate a point of general importance, not because I wish to give a chronology of current events.

The rules and procedures presented are current standard business practice, but they differ, sometimes dramatically, from older industrial engineering practice. These differences do not necessarily signify a gulf between the world of engineering and the world of business; rather, they highlight the contrast between the old and the new.

Technically trained readers, perhaps uncomfortable with modern management practices, may feel that I advocate decision making exclusively from a short-term financial perspective. It is true that one can damage an enterprise by making decisions based exclusively on a short-term perspective. But wise managers don't do that. They strike a balance.

Unfortunately, nonfinancial people sometimes ignore the impact of compound interest on the organization. They act as though time didn't matter, as though a dollar earned in five or ten years is as valuable as a dollar today. But at 14 percent interest, a five-year dollar is worth only 52 cents today, and a dollar to be paid in ten years is worth only 27 cents today, even if we neglect risk and the impact of inflation. Yet, as one midcareer engineer commented on Part II of the text, "It is not . . . an engineer's role to understand balance sheets and income statements. That's management's job."

Respectfully, I must disagree. Technical people, engineers, computer scientists, biologists, chemists, research and development personnel, and others employed in industry must know their specialist crafts, but failure to understand the fundamental financial basis of the enterprise risks rendering their managerial judgments irrelevant.

In this general, introductory survey it is impossible to treat each topic in depth. Entire courses and texts are devoted to topics considered here in a single chapter or section.

In the summer of 1981 I was privileged to attend "The Executive Program," TEP, of the Darden Graduate School of Business Administration at the University of Virginia, where I was introduced to many of the concepts discussed in this text. The faculty of TEP is outstanding, and their dedication and enthusiasm is infectious. I wish also to give special thanks to Dr. Robert W. House, Ingram Distinguished Professor and Director of the Management of Technology Program at Vanderbilt University, and to William F. Gibson for their careful reading of an earlier draft of the text and their many valuable criticisms and generous suggestions. I am grateful to the reviewers for Prentice-Hall whose names are unknown to me, to production guidance from Gretchen Chenenko, and the help of my graduate assistants, Richard Mathieu and Martin Sommer.

John E. Gibson
Ivy, VA

PART

I

The Strategic Business Environment

This first part provides a perspective of the environment faced by the manager of the modern American industrial enterprise. It is essential for professionals to understand the global factors affecting business organizations and some of the strategic issues general managers must consider to reach correct decisions. An understanding of these broad issues will help managers reach conclusions consistent with proper corporate strategy.

CHAPTER

1

Development of the Modern Industrial Enterprise

1.1 INTRODUCTION

The current strategic business environment faced in the United States by the private-sector manager might be described as a constrained free marketplace. While it might appear to be a contradiction in terms to use *constrained* and *free* in the same definition, in reality, there is no contradiction. The concept of a completely free marketplace is an abstraction that exists only in the minds of economic philosophers.

American entrepreneurs face currently and will continue to face in the coming decades, business challenges that are different in kind from those of the past. For example, American business enterprises are no longer able to consider global sales as merely a desirable add-on. Rather, for many firms global marketing is a vital necessity. Even those American concerns satisfied with their traditional national or regional sales are not safe, because international firms are invading the American marketplace with ever-increasing determination and vigor.

Managing for short-term results has become endemic in recent American business practice. Although there seems little doubt that orientation toward short-term results initially brought needed discipline to firms formerly content with sloppy or nonexistent financial and production controls, it has now exceeded the bounds of rationality, according to many observers. Indeed, this short-term attitude is a growing concern for leading business school faculty members, who feel responsible for initiating the trend and were among the first to detect its excesses. Part I of this text focuses on the longer-term planning horizon and the wider scope needed if American enterprise is to survive the international challenge it faces.

Contrary to the thesis of John Kenneth Galbraith in *The Affluent Society* [1], the United States has not solved the problem of efficient industrial production. Although the automatic factory is an isolated reality, rationalizing the reindustrialization of America with the status quo will provide many problems in the coming decades. Our grasp of industrial quality control remains generally inadequate, industrial productivity continues to advance too slowly, and industrial personnel relations remain for the most part bogged down in the morass of adversarial controversy. The second industrial revolution now under way in American industry has profound implications for management and labor in the United States.

1.2 HISTORICAL GROWTH PHASES IN AMERICAN PRIVATE ENTERPRISE

Before 1820

Because of the salutary sloth and inefficiency of eighteenth-century British civil service, as well as the great difficulty in transatlantic communication, the American colonies were allowed to develop mercantile practices more or less as they pleased. While tradespeople in the main colonial ports were harassed by taxes and tariffs, and Crown laws restricted products that could be shipped as well as the carriers that could be used, entrepreneurs in the hinterland were as free as even the most ardent modern libertarians could wish.

Thomas Jefferson, influenced by Locke and other thinkers of the Enlightenment, believed that government governs best which governs least. His thinking, and that of Madison and the other founding fathers, strongly supported free trade during and after the Revolution. Indeed, the American Revolution would never have claimed the adherence of the urban middle class had there not been royal interference with its private enterprise. Students of colonial and Revolutionary America will remember the Stamp Act, the Corn Laws, the various Whiskey Rebellions, and the Boston Tea Party.

In this period private ventures were usually sole proprietorships or partnerships. Great trading companies such as the East India Company did not succeed in America, although Pennsylvania and the Jamestown colony were initiated under this rubric. A few proprietors grew exceedingly wealthy, but the limitations of the owner-manager system caused most enterprises to remain small in America through the early decades of the nineteenth century.

1820–1850

Although new forms of management did not appear in this period, it did see the rise of substantial commercial enterprise and the opening of the upper Midwest. This was the era during which the City of New York established beyond all doubt its commercial superiority over its East Coast rivals Boston, Philadelphia, Baltimore, and Savannah. Robert Albion, in his fine book *The Rise of the New York Port* [2], shows that contrary to common opinion, this triumph did not come about naturally from an accident of geography, but, rather, the rise of New York can be traced to aggressive entrepreneurship.

The Pearl Street auction in New York City provided a smoothly functioning marketplace for wholesale traders. The Black Ball Line, the first scheduled sailing packet on the Atlantic, meant rapid, sure communication between New York and Liverpool. Because of alert New York brokers and factors, southern cotton growers soon found it more convenient to ship through New York than through Savannah, and, as Albion points out, by the 1850s approximately one-half of southern cotton revenues stayed in New York. Graded and branded New York flour sold at a premium over Philadelphia flour even in Philadelphia itself. The opening of the Erie Canal in 1825 did not cause the rise of the New York port but was instead a result. The canal reversed the flow of Great Lakes commerce, directing it eastward, and down the Hudson, whereas it had previously trickled westward toward the Ohio River and down the Mississippi.

1850–1870

The period immediately preceding and during the Civil War saw major changes in the way America did business. Prior to this period the typical small proprietorship could be managed by personal and informal methods, but now the increasing size of enterprises required a more formal approach.

One specific cause of change was the rapid development of the railroad. Prior to this period railroads were small, local concerns that operated at low speeds. In fact, one early railroad in Massachusetts operated trains moving eastward one day and westward the next. By the 1840s, however, utilizing more powerful locomotives over more carefully constructed roadbeds, railroads were capable of greater carrying capacity and speed. These engineering design improvements led to rapid expansion in railroad construction and "hard-driving" operation.

Unfortunately, this expansion resulted in an exponential increase in train wrecks and collisions. The solution resulted in a whole new way of organizing the industrial enterprise. It was no longer sufficient to have casual control of the workplace or finances. If discipline in the workplace were not established and maintained on a railroad, lives would be lost. And if strict control of the finances of a large commercial organization is not practiced, bankruptcy looms.

As Harold C. Livesay [3] notes, in the preindustrial America of the 1840s, if an individual employee moved too slowly or disobeyed orders, production on the farm or small workshop fell off in proportion. But on the railroad, in contrast, accuracy and promptness of operations was necessary to prevent catastrophe. Successful railroad executives soon learned these hard lessons.

> The Pennsylvania Railroad's management sorted out its labyrinth of financial and operating problems so effectively that it became known as "the standard railroad of the world." The principal architects of the bureaucratic structure making that achievement possible were J. Edgar Thomson, the railroad's first president, and Tom Scott, superintendent of its western division. [4]

Another pioneer of management was Daniel McCallum, general superintendent of the Erie Railroad. Some railroads of the period hired retired military officers as managers and adopted military rules of procedure. A few of these measures persisted—for example, the word *division*—but for the most part, railroad managers had to create anew.

Thomson of the "Pennsy" was a civil engineer as were Fink of the Norfolk and Western and most of the other pioneers, such as McCallum and Benjamin Latrobe of the B&O. A. D. Chandler, Jr., in his outstanding work of business history, *The Visible Hand* [5], speaks highly of the early contribution of American engineers and civil engineers, particularly as the first developers of many modern management principles. Chandler, Strauss Professor of Business History at the Harvard Business School, goes to the extreme of stating, "the pioneers of modern management . . . were all civil engineers."

Chandler cites Latrobe for developing financial accounting and operational precision and the creation of the first internal auditing staff. McCallum developed a number of principles of general management, and Thomson proposed the adoption of the line and staff concept. Perhaps the flavor of this creative period in American business management and the engineer's contribution can best be conveyed by quoting from *The Visible Hand*.

> There is little evidence that railroad managers copied military procedures. Indeed all evidence indicates that the answers came in response to immediate and pressing problems requiring the organization of men and machinery. They responded to these in much the same rational, analytic way as they solved mechanical problems of building a bridge or laying down a railroad. (p. 95)

> The resulting outcry [after several accidents in the 1840s] helped bring into being the first modern, carefully defined, internal organizational structure used by an American business enterprise. (p. 97)

> Certainly [McCallum developed] one of the earliest organizational charts in an American business enterprise. (p. 108)

> Of all the organizational innovators, J. Edgar Thomson and his associates [such as Tom Scott and Andrew Carnegie] on the Pennsylvania Railroad made the most significant contributions to accounting. . . . The new accounting practice fell into three categories: financial, capital, and cost accounting. (p. 109)

> The railroad manager who most effectively developed McCallum's proposals for cost accounting and control was Albert Fink, a civil engineer and bridge builder. (p. 116)

> The new methods, devised in the 1850s . . . remained the basic accounting techniques used by American business enterprises until well into the twentieth [century]. (p. 117)

> Indeed, the rise of American engineering education was, in part at least, a response to the needs of American railroads for trained civil and mechanical engineers. In the 1850s and the 1860s leading institutions of higher learning such as Harvard, Yale, Columbia, Pennsylvania, and Virginia offered specialized four-year courses in engineering. . . . These trained engineers . . . saw themselves and were recognized by others as a new and distinct business class—the first professional business managers in America. (p. 132)

Note the early and basic contribution of American engineers to the economic and managerial aspects of the industrial enterprise. Reciprocally, the centrality of such factors to the success of the engineering enterprise had been quickly and conclusively demonstrated to American engineers at the very beginning of modern American practice. Economics and management remained strong suits of college-trained engineers through the nineteenth century. The classic text on railway construction by A. M. Wellington, first published in 1877, provides an excellent example of this emphasis [6].

1870–1920

The period following the Civil War saw the development of major corporate enterprises in many important industrial sectors of America. Students of economic theory in this period saw before them evidence that, unrestrained by external forces, the free market economy tends to promote bigness and, ultimately, monopoly. Possibly in response to this observation, some political theorists turned to nonmarket economic systems such as socialism, while others advocated legislative constraints on the market system. "Trustbusting" became a favorite Washington pastime in that period, as environmental protectionism has in this one. In both cases regulatory excesses, if any, were brought on in large measure by avaricious and unprincipled business leaders acting in selfish disregard of the greater consequences of their corporate decisions.

Scientific management of the industrial enterprise in this period received added impetus from the work of Frederick Winslow Taylor. Taylor's concepts of industrial efficiency and scientific management methods are still widely used. Detailed consideration of Taylor and his contemporaries is deferred to Sec. 10.2 and the general implications of Taylorism and other management philosophies to subsequent sections of that chapter.

1920–Present

The current period can be described as the age of controlled or regulated free enterprise. As Galbraith wrote in the *Harvard Business Review*,

> Modern large-scale production, with its enormous investment and its long time horizons, would be impossible were prices, costs, and sales volume left to the uninhibited and wholly unpredictable movements of the classical free market. Equally, it is the purpose of farmers, oil producers, and workers to escape from the hazards and, on occasion, the cruelties of the market. The free and untrammeled market is greatly praised in the rhetoric; obeisance is rendered to it all over the industrial world; it is by no means dead. But the modern reality is a massive escape from its unpredictability. [7]

Accurate though Galbraith's perspective may be in global terms, it remains true that the United States starts with the concept of free enterprise in a market economy and modifies it through the government regulations Congress deems necessary. Many other nations, by contrast, start with the concept of a controlled and government-regulated economy, with freedom permitted when necessary and convenient.

The current period is also one in which many large American corporations face mature domestic markets, yet for most of these firms the world marketplace does not feel natural and comfortable. That is, these firms conceive of themselves as American business enterprises that operate internationally rather than as transnational corporations that operate in the United States and in other countries. Perhaps as a consequence of this parochial view, increasing numbers of U.S. firms have retreated to focus on short-term adjustments rather than engage in the creation of new wealth through innovative industrial production and aggressive international marketing.

The evolution of modern management style is summarized in Table 1.1.

TABLE 1.1 MANAGEMENT STYLE AND GOVERNMENT STANCE IN MAJOR PERIODS OF U.S. BUSINESS HISTORY

Period	Management Style	Government Stance
Before 1820	Sole proprietorships, guild system, indentured servants	Britain makes rules for colonies, lethargic bureaucracy
1820–1850	Same	Untrammeled free enterprise
1850–1870	Emergence of corporate entities, led by railroads	Same
1870–1920	Growth of corporations, predatory capitalism; Taylorism	Emergence of need for regulation of marketplace, trustbusting
1920–present	Focus on short-term, balance-sheet games	Alternating regulation and relaxation, mixed approach

1.3 GROWTH AND MATURITY OF THE INDUSTRIAL ENTERPRISE

The American automobile industry is a paradigm of the growth and maturation pattern of an industrial enterprise. Contrary to the popular view, the modern gasoline-powered automobile was not invented in the United States and Henry Ford was not its progenitor. At the turn of the century the concept of the automobile was in the air. It was a common topic of conversation, and many young entrepreneurs decided to make the automobile the tool of their business success. For a modern analogy, consider the rapid development of the small computer and commercial products based on solid-state electronics.

In the early 1900s no consensus existed in the U.S. automobile industry on what the customer wanted, nor even broad agreement on the technical means of achieving a practical motor vehicle. In 1902, to pick an arbitrary year in the period we are discussing, there were 106 steam car manufacturers, 40 electric car makers, and 99 gasoline-powered vehicle builders. Most of these entrepreneurs were small and underfinanced, and none had any exclusive technical advantage not available to the competition. The competitors were united in one thing, namely their common desire to achieve financial success in this exciting new field.

From this primordial technological ooze emerged several likely candidates for success. Now, perhaps, it is obvious that the market demanded an economical, simple, rugged people's car such as the Model T. But no such surety was given to the 245 striving contenders in 1902. In fact, the earliest believer in this concept of the "people's car" met an interesting fate. Ransome E. Olds had been designing and building electric and gasoline-powered automobiles since 1886. In 1892 he founded the Olds Motor Works in Detroit with financing from a local copper magnate. The venture was at best marginally successful as Olds vacillated on questions of price, markets, and motive power. Finally, in 1901, close to bankruptcy, he hit the mother lode with the "Curved Dash Runabout." It was light, simple, and cheap at $650. Olds sold only for cash, the universal practice at the time, but he did discover through necessity one element of future standard automobile industry management practice.

Olds had no money for giant factories and huge payrolls. Thus he was forced into extensive subcontracting. Because of the hundreds of tiny machine shops in the Detroit urban complex, Olds was able to find original equipment manufacturers (OEMs) who would construct to his design all of the major parts for the runabout. He was the designer, the assembler, and the distributor. Immediate success followed announcement and exhibit of the runabout at the New York Auto Show in January 1901, and Olds sold 425 units for the year. Figure 1.1 shows the sales of the Curved Dash Runabout for its entire lifetime. Note the approximately linear increase in sales as the price is held constant, this seems to be a sales performance characteristic of a product facing unfilled market demand.

It is interesting that, to some managers, preconceptions and theories are more convincing than facts. This was true for the owners of Olds Motor Works. Despite the success so obviously enjoyed by their low-cost model, the owners insisted that the company's future lay with larger, heavier, more expensive cars. Imagine the ancestors of modern-day auto industry managers saying, "We can't make a profit at the low end" or "There's no money in compact cars." Finally, Olds grew tired of corporate infighting and withdrew from the company in 1905. Soon thereafter, production of the runabout was canceled and the company withered. Several years passed before Henry Ford stopped production of his marginally successful, expensive models and introduced the low-cost Model T.

Late in 1908 Henry Ford introduced the Model T at a basic sales price of about $900. Figure 1.2 shows the remarkable sales growth of this venture. Note the exponential increase in sales as the sales price is continuously decreased, characteristic of products facing unfilled market demand. The product's value in use to the consumer exceeds the purchase price and creates market pull. (Compare with Fig. 1.1.)

It is clear that Ford correctly sensed the doctrine of the market. His product embodied what customers wanted better than the competition's, and Ford's reward is obvious from the sales graph. Ford was determined to build the "farmer's best friend." He wanted a light, simple, cheap, easily repaired source of basic transportation and power that could be operated over country roads in all types of weather. Ford visualized the Model T being repaired by the farmer in his own barn and with his own tools. He

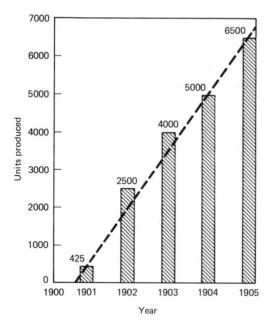

Figure 1.1 Lifetime production of the Olds Curved Dash Runabout—sales price $650.

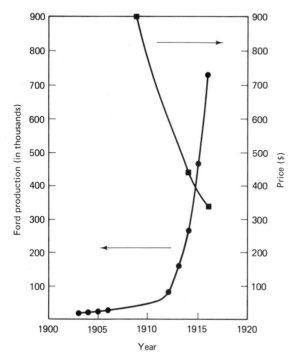

Figure 1.2 Ford production versus price. Model T production began in late 1908.

foresaw the Model T with its wheels off and up on blocks, acting as a donkey engine to power hoists, sawmills, and the like.

The foregoing is the well-known romance of the Model T, but modern managers have more than romance to learn from Ford. For instance, why was Ford successful with his people's car when others such as Olds weren't? Ford wasn't first with the concept as we know, but he was certainly more determined. In fact, *determined* is inadequate to describe Ford's fanaticism; *maniacal* might be a better term. Ford was determined to let nothing stand in his way. He reorganized his company to force out stockholders who resisted his single-minded focus on vigorous price reduction and rapid growth financed from operating surplus. Ford poured back into the company almost all of its earnings so as to expand production. Against the vote of a majority of the stockholders and a court order he was still reluctant to increase dividends.

Ford was fanatically determined to cut costs. Today if a manager cuts prices below cost to run down the learning curve more rapidly (see Sec. 2.2), he is considered bold and risk-seeking. Imagine the risk to Ford when he invented this concept. Modern Japanese managers sometimes anticipate the cost reductions achievable from the learning curve, but Henry Ford did it first.

Note from Fig. 1.2 that Ford was gaining more and more sales and other less successful manufacturers must have resented his "predatory pricing." Ford was not merely "stealing sales" from other makers. He was creating new sales! New categories of automobile buyers were brought into the market by Ford's price cutting, and his aggressive marketing style forced other manufacturers to become innovative to stay alive.

Most products exhibit what economists call price elasticity. That is, as the price of an object is decreased, it becomes affordable to more and more potential customers and total sales increase. Thus price cutting by Henry Ford did much more than take sales from other manufacturers. It greatly enlarged the total market and made it possible for whole new population sectors to benefit from owning an automobile. Enlarging the market could have indirectly aided Ford's competitors if they had responded to his challenge. This process is economic Darwinism: survival of the fittest.

As will be explained more fully in Sec. 2.2, once Ford had achieved the dominant share of the market, his costs fell below those of his competitors. Why then did he not go on to achieve a complete monopoly? In theory this is the logical outcome of the simplified process just described, assuming that new factors do not intervene. In practice, other factors almost always do intervene, and the sales curve saturates before complete monopoly is achieved.

Figure 1.3 shows the generic product life cycle (PLC) curve. There is slow growth initially, then sales take off in a rapid exponential curve, provided there is unmet market demand and the price is reduced as production costs go down. Finally the curve saturates as unmet or pent-up demand is satisfied. Product sales enter the mature or slow-growth region. Maturity may continue for decades, as with Hershey's chocolates, for example. Some products exhibit multiple plateaus—new and improved washday miracles, for example—while other product sales curves dive more rapidly than they first rose—hula hoops, for example.

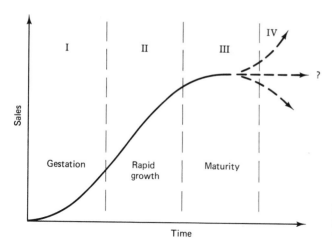

Figure 1.3   The usual *S*-shaped growth and maturity curve of a successful product innovation (the generic product life cycle curve).

1.4 FACTORS INHIBITING GROWTH OF A POTENTIAL MONOPOLY

Three major factors inhibit the growth of a potential business monopoly. First is a group of so-called structural factors, such as market fragmentation. Second is threatened or actual government intervention, and the final and least understood factor is the inevitable change in market doctrine.

A potential monopoly is an enterprise that has an important advantage over its competitors because of a functionally superior product and a more advantageous cost situation, possibly because of a dominant market share and consequent experience curve position. It should be emphasized that this competitive advantage rarely, if ever, develops by accident; it is the result of good management. The directors of the dominant enterprise may choose not to take advantage of a potential monopoly for their own reasons. But these are not the only reasons. The owners of the Olds Motor Works chose not to take advantage of their position with the Curved Dash Runabout in 1905, and this choice led to incipient financial failure and forced liquidation.

You need not go back as far into ancient automotive history as 1905, however, to discover a failure to take advantage of a potential monopoly. British Leyland's failure to market the Land Rover properly provides a more recent illustration. The Land Rover was a unique vehicle and had no competition for years after its phenomenally successful introduction. But BL management seemed deliberately to ignore the opportunity to dominate the off-road, four-wheel-drive market sector worldwide. The *Wall Street Journal* put it as follows:

The Land Rover's exceptional success is anything but a triumph of skillful marketing, as its current makers readily concede. That the vehicle exists at all today is largely accidental. . . .

In the 1960s and 1970s, the market for four-wheel-drive vehicles did the opposite of what Land Rover's people predicted. It grew. By 1980 it had reached 850,000 cars per year. But Land Rover continued to produce at its historic level [of 50,000 units per year]. [8]

But there can be valid reasons for restraining an organization to less than complete monopoly of a mature market. Pierre Du Pont, when he commanded the Du Pont company in the first few decades of the twentieth century, believed in limiting one's share of a mature market to approximately 60 percent. This share permits a company to be the dominant, low-cost producer and thus the market price leader. In bad times, the low-cost producer can cut prices to maintain full production while smaller, and thus higher-cost producers lose sales and possibly go bankrupt.

Market structure elements to prevent a potentially dominant producer from achieving a monopoly include the following factors. Suppose a product's distribution cost varies with the distance from the plant to the customer, and this distribution cost is a significant portion of the total product cost. In this case a company will not be able to ride down the experience curve (Sec. 2.2) to achieve a monopoly because a competitor can open a plant closer to distant markets and have lower transportation costs. An example of this situation is the retail distribution of baked goods, milk, and other perishables.

A venture may be limited by the availability of raw materials and thus be unable to increase production. The market may be fragmented, making it impossible to achieve economies of scale. A business may not be able to protect the initial functional advantage of its product over the competition, common in the fashion industry. Other examples of structural limitations will also occur to the reader.

Government regulations may prevent achievement of a monopoly. American society considers a monopoly bad because it permits the monopolist to set prices and gouge the customer. Sufficient examples in business history exist to make politicians unlikely to challenge this popular opinion. For many years General Motors (GM) felt that if it exceeded a 50 percent share of the U.S. automobile market, it ran the grave risk of antitrust action. Thus it voluntarily restricted its U.S. sales. As the dominant, and low-cost auto producer prior to and immediately following World War II, GM enjoyed approximately double the rate of return of other U.S. auto manufacturers and could have put smaller rivals in a severe price squeeze at any time. Sometimes the dominant producer fails to notice the changing market doctrine and thus is displaced even while attempting to monpolize the total market. This happened to Henry Ford in the 1920s.

1.5 THE AUTO MARKET DOCTRINE CHANGES

By 1916 the sales growth of the Model T slowed to a handsome 30 percent annually and continued to hover about this number, with wide annual variations it is true, until 1924. In the lexicon of the Boston Consulting Group (BCG), Ford was no longer a "shooting star" but had settled into the "cash cow" category. BCG classifications are discussed in Sec. 2.3. In the meantime, another actor was preparing for a major challenge: General

Motors. In 1916 Ford produced almost three-quarters of a million Model Ts alone, while GM's total production for all units, including trucks, was barely 20 percent of Ford's. But ten years later, in 1926, GM sales equaled Ford's, and never again would Ford sales top GM's. What happened?

Henry Ford provided the market with the basic transportation he believed people needed. The market agreed with Ford, and the Model T was on its way. But Ford did not decide for the people what they wanted. The market decided for itself, and the Model T was consistent with its choice. Ford never understood this distinction. He made a mistake repeated by many modern American managers when he thought that he could dictate to the market what it wanted.

Auto magnates, untutored in economics and the theory of the market, may perhaps be forgiven this error, but even the well-trained economist John Kenneth Galbraith fell into the subtle trap of confusing synchronicity with cause and effect. In 1958 Galbraith postulated that U.S. automobile companies could independently determine price and style and force their choice on the American consumer [1]. No doubt he has since changed his mind.

In the 1920s Alfred P. Sloan, Jr., took charge of GM and established a new market doctrine that held until the 1970s. Unlike Ford, Sloan did not found his company but was brought into the newly formed GM by William C. Durant, a dreamer who created GM by buying smaller automobile companies and parts suppliers. Durant's fanaticism for growth by acquisition matched Ford's mania for internal growth. However, Durant's position in GM was not as secure as Ford's. Durant needed external financing, and when he overextended himself financially in 1910, he was forced by his creditors to give up control of the corporation.

Durant immediately started an independent enterprise, Chevrolet Motors. In 1916 he used Chevrolet assets and funding from Du Pont to regain control of GM. In the 1920 depression, the auto business collapsed, and Durant's futile tactic—holding firm on prices in the face of a 20 to 30 percent price cut by Ford—caused a sharp drop in GM sales, leading Du Pont and others to remove Durant from control of the corporation for a second and final time. Du Pont turned to the grim, gaunt Sloan to provide active corporate leadership at GM.

Sloan's book, *My Years with General Motors* [9], is a classic on corporate management. Sloan tells how he came to believe the auto buying public was ready for a new doctrine that came to be called trading up. More generally, we would now classify Sloan's new definition of marketing as an example of *market pull,* which should be contrasted with Ford's style of deciding unilaterally what the consumer needs, and is now called *technology push.* Sloan argued that the new auto market should feature the following four elements:

1. Consumer credit
2. Used car trade-in allowance
3. Closed (all-weather) body
4. Annual model change

The annual model change had been a feature of the early auto business but was dropped before World War I. Ford fought each of Sloan's marketing innovations and the functional modifications forced by the annual model change. But Sloan had correctly anticipated the new market, and as a result, GM became the marketing and functional-feature leader of the period. In 1925 Ford began a decade of losses that by 1935 amounted to over $26 million.

What lessons concerning the strategic business environment can be learned from this specific example of the loss of leadership by Ford? It is true that the failure at Ford was more dramatic than most, and by 1926 Ford would have been removed by the board of directors of almost any publicly held company. But there are many nonautomotive examples of failure of a market doctrine and refusal by a crumbling market leader to face reality.

Durant was not allowed to run amok as did Ford, and this was to the benefit of almost all concerned. Of course, had Ford and Durant not been the fanatics they were, perhaps neither would have run the risks they did. Neither Ford nor Durant was a manager in the modern sense. Both were primal creatures bent on creation and driven by a wild sort of vision remote from economic reality. Edwin Land and his insistence on developing personal hobbies at the modern Polaroid Corporation, even though it is a publicly held firm, provide a more recent example. Among the 240 early auto entrepreneurs there surely existed a number of good managers, but they weren't allowed to stay in business long enough to demonstrate their more modest talents.

1.6 THE AUTO MARKET MATURES

By the late 1930s the auto market was mature. The days of 30 to 50 percent growth per annum were gone forever. The U.S. auto market would grow with the gross national product. In good years sales advanced more rapidly and in poor years they slumped badly, because a new automobile is a discretionary purchase. Ford slipped to third place behind Chrysler, a new organization bent on leadership through technical advances, while GM was clearly the dominant producer. The Sloan market doctrine was successful and static. All technical advances had been made, and the product took on a commodity-like appearance. Automobile advertisements began to emphasize ancillary benefits such as styling, color, and peer approval because all vehicles were functionally equivalent.

It is a remarkable tribute to the power of a marketing image created by clever advertising that automobiles continued to play an important role in the life style of Americans until the early 1970s, long after the automobile had become in fact a commodity. In principle, the automobile could have slipped from the American consciousness in the 1930s as did other useful but utilitarian artifacts of modern life when they reached maturity. But by dint of carefully crafted advertising and the willing cooperation of the public, the American auto continued to be a glamour accessory for 40 years.

By the 1970s, however, as a *Wall Street Journal* headline noted, "People Are Keeping Cars Longer As Costs Rise and Attitudes Change" [10]. The average age of autos registered in the United States has increased by more than a year in the 1980s (resulting in approximately $50 billion in lost sales). People are driving fewer miles annually and running their cars more total miles before trading. Higher first costs and operating costs play an important role in the decision to retain autos longer, but, as the *Wall Street Journal* notes, it is also true that attitudes have changed. Social status is no longer conferred by ownership of a new car. Why is this so? Why did it happen only with the past few years rather than forty years ago?

Modern automobile magnates have contributed to this lost cachet by blurring sales segments as well as by failing to control costs. Perhaps modern American auto leaders, with a few exceptions such as Lido A. Iacocca, simply do not understand their business. Like Henry Ford, they have unthinkingly continued with the market doctrine of the past, long after it ceased to be viable. Furthermore, they focus on short-term economic return while ignoring the approaching crisis.

The style and personality type of managers in a mature industry differ from the personality type dominant in a growth industry. As a class, managers in the auto industry began to change as the auto business matured. They grew older and more conservative. They distanced themselves from the factory floor, and rather than taking pride in new technical achievements, they grew concerned with other things. Talent at financial control and office politics became more important than attempting to plan ahead for market changes.

Figure 1.3 shows several dotted trajectories for the product life cycle after the product maturity plateau is reached. If nothing disturbs the equilibrium, sales can continue at a high level indefinitely. But the longer this stasis continues, the more likely a major discontinuity becomes. Sales may decline as the product is replaced by something closer in tune with the evolving market doctrine, or sales may increase if the product is modified to meet or anticipate the market.

In the automobile industry some product changes did take place between 1935 and 1975. Automatic transmissions, air conditioning, more reliable tires, sealed-beam headlamps, push-button radios, two-tone paint, and a windshield wiper for the passenger side are among the advances to be noted. Many of these changes are of minor import, such as a disappearing windshield wiper and a fraction-of-an-inch increase in axle length. The latter change was an advance sufficient to cause the city of Pontiac, Michigan, to rename its main boulevard ''Wide-Track Drive.'' Yet as we have noted, such cosmetic and advertising maneuvers kept the automobile in a mature high-sales state for decades longer than a more reasoned approach might have.

The American automobile industry was mature for over 30 years, and in 30 years several generations of general managers ruled in succession. In a mature industry the ''dwell time'' of the top level of management is reduced to as little as five years, because reducing tenure at the top is the only way to hold on to professional middle-level managers. If the company cannot offer a shot at the top to its aggressive M.B.A.'s, they will move on. U.S. automakers were less influenced by fads in business administration than many large firms, but, especially at GM, the corporate headquarters group distanced itself from what might be called a strategic viewpoint of the automobile and its place in the mix of transportation modes. The group focused ever more narrowly on itself and the "styling game." Life at GM headquarters was not very different from the world described by Hermann Hesse in *The Glass Bead Game* [11].

The culture of the status quo became deeply embedded in Detroit. Those few people who attempted to challenge the status quo from within were isolated and driven out. John Z. De Lorean is an example of the brilliant, hard-driving manager who found it difficult to play the corporate styling game [12]. Had the auto industry allowed itself to rethink the Sloan doctrine as it came under increasing attack and decentralize, it might have been able to retain the mavericks such as De Lorean for a time when they would be needed. But when the challenge of change comes, the rigid, dogmatic, ingrown manager of a mature industry is particularly ill-equipped to cope.

1.7 LE DÉLUGE

Many students of the American automobile industry became concerned in the 1960s about the increasing likelihood of cataclysmic change. The popularity of the Volkswagen and other foreign vehicles among social style leaders in the United States signaled the approaching discontinuity to those who wished to take note. The discontinuity could have taken many forms, including the following possibilities.

Vehicle Technology.   Suppose the electric car had proved technically and economically feasible in the late 1960s or early 1970s, or suppose Lear really had developed a practical steam engine. Would the existing American automobile manufacturers have been able to respond to this technical discontinuity, or would they have resisted change until it was too late?

Production Technology.   The automatic factory has been possible in theory for 20 years, and it has been a reality in Japan for five. The U.S. automobile manufacturers' failure to adopt automation more rapidly has contributed to high labor costs and low product quality.

Distribution Innovations.   Suppose it became practical to assemble autos with only a dozen or so workers. Such an approach could be located in any major automobile dealer's repair facilities, and many automobile assembly plants could be eliminated. Would the American auto companies adopt such a change?

U.S. auto leaders had little difficulty persuading themselves that extravagant labor costs could be passed on to the captive consumer, and the brilliant leadership of the United Auto Workers Union pressed this advantage. Management did nothing to remove itself from the trap and compounded the problem by contributing to worker alienation in its cynicism. By excessive model proliferation and blurring of market segments, management lost the product definition it had spent decades and hundreds of millions of advertising dollars to establish.

The disastrous loss of market share of Chevrolet in the period from 1975 through 1981 is a clear example of this marketing failure [13]. Chevrolet's image was blurred and its price advantage lost. As a result, it slipped from approximately 50 percent of total GM sales prior to 1975 to less than 38 percent in 1981, and total GM sales were falling in this period as well. Some critics argue that by the 1960s and 1970s the leaders of the American automobile industry would have been helpless in the face of any of the discontinuities just mentioned. To these critics it was only a matter of chance that the actual threat came from overseas (see Fig. 1.4). As Fig. 1.4 demonstrates, domestic production is approximately a constant share of a declining market. The market declines because an automobile is an optional purchase in the short run. Sales of Japanese cars are not decreasing in the United States.

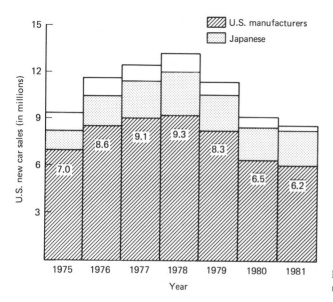

Figure 1.4 U.S. automobile sales by domestic and foreign manufacturers.

1.8 IS THE AUTO STORY RELEVANT?

The story of the automobile industry is interesting, but does it provide any general strategic managerial insights? The sales trajectory of Ford in the 1930s and of the American automobile industry as a whole in the 1980s conforms to a standard scenario worked out by the Boston Consulting Group, which is discussed in Chapter 2. But before that general formulation, consider a more modern example.

One problem with a discussion of the automobile industry is that it may appear stale and irrelevant because it is rooted in a technology almost 100 years old. Of course, people who refuse to learn the lessons of history may be doomed to repeat them, but it is not necessary to go back into history to discover a new business nearing the end of an explosive growth phase and based on newly developed technology that meets a need in the marketplace.

In 1982 there were more than 150 new companies focusing on personal computers and dozens of new ones being formed every month. Word processing, small business applications, computer games, and interactive graphics for engineering design are just a few of the applications. Many of the most successful entrepreneur-founders of these organizations were young. A number of computer millionaires were still in their twenties or thirties. Innovative hardware, new software, and new marketing efforts met with incredible success. Young engineers and business majors trekked to the Silicon Valley to earn their fortunes. After a few years learning the business, many went off on their own to found new companies. Venture capitalists were eager to provide funding to young computer entrepreneurs. Wall Street took computer companies public, and the stock skyrocketed before the company made a profit, sometimes before it even had a product.

The contrast with the modern auto business is apparent. Yet if you look closely, both industries are following the same clearly defined growth track. In fact, the only difference in the scenarios outlined for the microcomputer industry and the automobile industry is that Henry Ford's growth was more rapid than the growth of all the personal computer companies put together. Figure 1.5 shows the two sales curves superimposed for comparison. It may surprise you to see that Model T sales grew at a faster pace than personal computer sales. The rapid growth in Model T sales is all the more remarkable given the smaller U.S. population of the period: 102 million in 1914, 235 million in 1983.

As in the early automobile industry, new computer products are the result of extensive subcontracting, with many different end products using the same integrated-

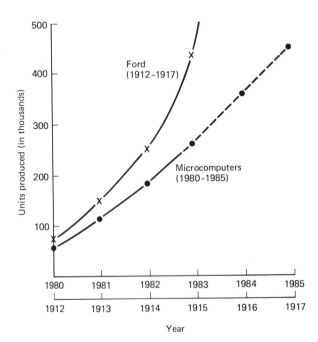

Figure 1.5 Recent past sales growth curve of personal computers superimposed on the Model T sales curve.

circuit components. Computer manufacturers take in each other's washing by subcontracting components for each other and sometimes owning stock in each other's ventures. Just as the bar at the old Hotel Pontchartrain in downtown Detroit became the social center for young auto magnates in the early 1900s, watering holes in Cupertino and Palo Alto now serve the same purpose.

As in the early automobile industry, the early 1980s shakeout in the personal computer market was hard to survive. Many young computer entrepreneurs, such as Ohio Scientific and Osborne, were forced to sell out or declare bankruptcy. IBM restructured the PC marketplace when it entered in 1980. By 1983 IBM had taken over market leadership from Apple, just as GM had surpassed Ford. In 1985 the PC marketplace was again restructured by the sales growth of cheap, close copies of IBM's PC, called clones. If Henry Ford, R.E. Olds, Billy Durant, and the other early automotive pioneers could be brought back and given a few days' briefing on the technical details, they would find the computer business all too familiar.

Anecdotes and history serve a useful purpose, but now we need to establish a solid theoretical base for the standard product life cycle and use it to discuss the standard entrepreneurial approach of the Japanese in advancing worldwide in the high-technology marketplace. We should then be able to predict the outcome of the global electronics business. All of this we do in Chapter 2.

EXERCISES

1. What is a commodity-like product?
2. Why did Pierre Du Pont set his desired market share at 60 percent?
3. Did Henry Ford invent the automobile? The low-cost automobile?
4. Did the Erie Canal cause the rise of the New York port?
5. What was Henry Ford's market doctrine?
6. What was Alfred P. Sloan's market doctrine?
7. What would you propose as a modern automobile market doctrine?
8. What is meant by the term *price elasticity?*
9. In Fig. 1.2 the sales growth of the Model T is said to be typical of unmet market demand and a demonstration of *market pull,* yet near the end of Sec. 1.5 Ford's marketing philosophy is said to represent *technology push.* Explain these terms, and resolve this apparent contradiction.

REFERENCES

1. John Kenneth Galbraith, *The Affluent Society* (Boston: Houghton Mifflin, 1958).
2. R. G. Albion, *The Rise of the New York Port* (New York: Scribners, 1959).
3. H. C. Livesay, *Andrew Carnegie and the Rise of Big Business* (Boston: Little, Brown, 1975).

4. Ibid., p. 36.

5. A. D. Chandler, Jr., *The Visible Hand: The Management Revolution in American Business* (Cambridge, Mass.: Harvard University Press, 1977).

6. A. M. Wellington, *The Economic Theory of Railway Location,* 6th ed. (New York: Wiley, 1899).

7. John Kenneth Galbraith, ''The Way Up from Reagan Economics,'' *Harvard Business Review,* 12 (July-August 1982), 6–8.

8. B. Newman, ''Swamps Don't Stop British Land Rovers, but a Strike Might,'' *Wall Street Journal,* October 30, 1981, p. 1. (Reprinted by permission of *Wall Street Journal,* © Dow Jones & Company, Inc., 1981. All rights reserved.)

9. A. P. Sloan, Jr., *My Years with General Motors* (Garden City, N.Y.: Doubleday, 1964).

10. C. W. Stevens, ''People Are Keeping Cars Longer as Costs Rise and Attitudes Change,'' *Wall Street Journal,* January 7, 1982, p. 23.

11. Hermann Hesse, *The Glass Bead Game* (New York: Bantam, 1970).

12. J. P. Wright, *On a Clear Day You Can See General Motors* (New York: Avon, 1979).

13. R. L. Simison, ''Skidding Pacesetter: Chevy, GM's Leader, Sustains Worst Slump in U.S. Auto History,'' *Wall Street Journal,* January 4, 1982, p. 1.

CHAPTER

2

The Current Strategic Environment

2.1 INTRODUCTION

Chapter 1 briefly discussed the past development of industrial America using the auto industry as an example. Chapter 2 focuses on current strategic patterns in American industry. The current U.S. strategic business environment is dominated by the need to react effectively to intense foreign competition. In past eras American manufacturers were content serving a regional or at most the U.S. national market, but that is no longer the case. American industrial managers who find the export trade too strange and complex for their organizations and are willing to settle for serving U.S. markets will not be let alone.

Foreign competitors find it necessary to enter the U.S. internal market, because their internal markets are generally too small to sustain production runs large enough to permit economies of scale. Neither Germany nor Japan, perhaps the most intense rivals of the United States, could sustain their current standard of living if forced to cut off exports, and they have no intention of trying.

Three specific elements in the overall response of U.S. industry to this international challenge are of particular interest. First, the development of the conglomerate concept upset traditional industrial patterns, but it was initiated by Royal Little of Textron in direct response to the Japanese industry's threat to Textron's traditional textile business. Second, some large high-technology organizations have learned to manage themselves without becoming conglomerates, and the distinction is important. Finally, Japan's development of a national industrial policy, led by its federal Ministry for International Trade and Industry (MITI) and focused on global economic conquest, will be examined.

The U.S. auto industry did not succumb to internal competition but instead to a combination of international forces. Furthermore, many American businesses are subjected to the same form of international competition that threatens the continued existence of the U.S. auto industry. Cameras, electronic items, motorcycles, watches, computers, photocopy machines, lawn mowers—all of these items and others have been threatened or displaced by foreign competition. From a study of the phenomenon of international competition in several industries a clear pattern emerges. The existence and understanding of this pattern means industry in the United States should be able to predict the next moves and reposition itself to anticipate them. Unfortunately, repositioning may be easier said than done.

Examining the global marketing strategy used by Japanese industries and the reactions of U.S. firms, it is evident that, contrary to folk tales, the Japanese are not winning by using low-cost, docile labor, by dumping goods onto international markets at a loss, by cheating on agreements, or by bribery, although they have used all these techniques in the past and will use them again in the future, just as some U.S. firms have.

Japanese firms have studied well-known American industrial enterpreneurs and are using their proven techniques to help defeat American successors. Furthermore, success in the international marketplace will be increasingly required of American corporations because further growth in American markets will be limited for many products. Finally, short-term optimization of corporate income can be detrimental to the firm's long-term health if pursued in a single-minded fashion.

2.2 COST REDUCTION AND THE LEARNING CURVE

In almost all well-managed industrial production operations it is possible to reduce direct production costs for a process that continues in use over a period of time. This cost reduction occurs for a variety of reasons and is described by the term *learning curve* or *experience curve*. The first and simplest reason for cost reduction is that workers become more skilled and produce more for an hour's effort, although this factor is not as important as nonexperts may think. Second, as production engineers become more familiar with the process, they can generally design shortcuts and simplifications, especially if they are willing to listen to operating employees' suggestions. Finally, design engineers can often redesign the product to simplify it and thus reduce manufacturing costs. Although economies of scale can be considered separate phenomena, they come into play as well.

Henry Ford was perhaps the first to recognize intuitively the economic power inherent in the learning curve. He believed that as an organization gains experience manufacturing a product, its manufacturing costs should come down. The simplest example is that of a single worker gaining skill and speed with practice—but the learning curve is more. An auto assembly worker can learn all there is to know about any job on the assembly line in a few days, yet Abernathy tells us the Model T rode down an 85 percent learning curve for almost 20 years, from 1908 till 1926 [1]. An 85 percent

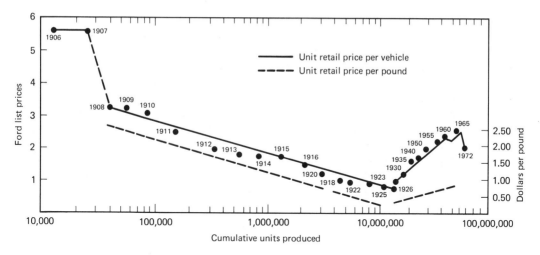

Figure 2.1   The learning curve for the Model T. (After Abernathy and Wayne [1])

learning curve means that each time total production is doubled, unit production costs are reduced to 85 percent of their previous level.

Careful attention to production details permitted Henry Ford to reduce Model T costs continually for almost 20 years. The graph of cost reduction often appears linear on a logarithmic scale, as in Fig. 2.1. Note that production cost is plotted as a function of total production and not against time. Model T production was discontinued in 1926.

There is nothing automatic about the learning curve. Top-notch manufacturing engineers and production managers must give constant attention to the processes involved. Production methods must be simplified and parts reengineered to make them simpler. Investment in labor-saving machines and material-handling devices is necessary. If an enterprise can achieve the benefits of cost reduction promised by the experience curve concept, we can see how powerful an incentive rapid sales growth can be. Higher sales mean that the investment needed for automated production methods can be shared across a higher base, and this sharing further reduces unit costs. If our market share is greater than the competition's, then our doubling time is shorter and we can cut prices to gain a still higher share while still earning the same profit as our higher-cost competitors.

Although the learning curve concept was known intuitively by Henry Ford in the early 1900s and rediscovered by the Air Force and Boeing in World War II, the concept has been put to its fullest use only recently, in the past 10 or 15 years, by management consultants such as the Boston Consulting Group. In its broadest form, the learning curve has more recently been called the "experience curve."

It is easy to understand the emphasis placed on market share by general managers who have been indoctrinated by BCG. These managers see that reduced direct costs brought about by the workings of the learning curve provide great flexibility. As costs are reduced, a company can lower the selling price while retaining the same profit margin, or the price can be kept high and extra profits captured. Thus it is imperative to move down the learning curve more rapidly than one's competitors if one is to stay in control of the market. Control is accomplished by maintaining a larger share of the market than the competition.

On occasion one hears of "anticipating the learning curve." This means that a producer sets prices lower than would be justified by current volume in order to gain market share; the larger share, if obtained, results in lower costs. This is also called "betting on the come," and other less polite names, such as "predatory pricing," and "restraint of trade." One reason the Douglas Aircraft Company is no longer an independent organization is that it anticipated in vain cost reductions to be produced by a rapid learning curve on the DC-9. High turnover in the southern California labor market prevented Douglas from realizing anticipated savings and brought the company close to bankruptcy. McDonnell Aircraft was then able to buy Douglas out.

BCG believers argue that if the dominant market share and thus low-cost producer position is lost there can be no recovery, provided the new dominant producer stays alert. The Boston Consulting Group doctrine dictates that once share starts to slip, if prompt action doesn't cure the problem, then the product and the division producing it must be promptly sent to the sales auction block.

In Chapter 1 we saw that British Leyland (BL) failed to consider both the concept of the learning curve and the company's dominant position as the producer of the Land Rover. Thus Toyota was able to drive BL to the wall. Some analysts argue that the Land Rover division of BL always made money in its nice little market niche, thus negating the argument. But BL as a whole went bankrupt. Unless the corporation as a whole maximizes return from its profitable divisions, it won't have the funds necessary to start new enterprises and reinvest for growth.

A firm must not allow a competitor to outdistance it in total production if the learning curve concept is valid. The dominant supplier can choose to maintain the product price, "keep up the price umbrella," and protect smaller producers while earning greater unit profits, as General Motors did in the 1940s and 1950s. Alternatively, it can choose to drive the price down, "collapse the price umbrella," and thus drive higher-cost producers to a loss position and ultimate bankruptcy, as Ford did early in the Model T's life cycle. But relinquishing the dominant market share position, as British Leyland did with its Land Rover, delivers the company's fate into the hands of the competition.

The life cycle of a product, the PLC, and how a product first extracts growth funds from the corporation and then can be made to repay this investment has been the subject of study by the Boston Consulting Group and is discussed next.

2.3 QUESTION MARKS, STARS, COWS, AND DOGS

The sales trajectory of the Ford Model T 70 years ago and of the American auto business as a whole in recent years conforms to a standard scenario worked out in the colorful terms at the head of this section by the Boston Consulting Group. The BCG schema we are about to discuss shows that market share and the rate of sales growth are remarkably sensitive predictive tools in strategic marketing, much more so than older, more conventional measures such as gross sales or profit. Had the firms to be discussed in the applications of the BCG matrix been sensitive to the indications of this important predictor, American business history might have been quite different.

In one sense, perhaps, there is nothing new in this BCG schema because the elements are familiar, but BCG must be credited with assembling the parts into a coherent theoretical whole. The BCG approach utilizes a four-element matrix shown in Fig. 2.2. Along the horizontal axis is plotted the product's market share, expressed as a fraction of the largest competitor's share. Note that on the horizontal axis, relative market share increases to the left, which may seem a little confusing at first. The center of the horizontal axis represents the point at which our sales in this market segment equal the sales of our largest competitor. The vertical axis represents our annual sales growth, expressed as a percentage of our sales in the previous year. The center of the vertical axis might be 10 percent growth, with the bottom representing zero growth and the top 20 percent growth. The vertical scale is chosen to suit the situation at hand. Looking forward to Fig. 2.4, for example, the sales of the Ford Model T are charted with respect to total General Motors sales in the same year. In that chart a vertical axis that displays Model T sales from zero to one hundred percent annual increase was required.

Lets see how the BCG matrix works in the abstract before applying it to Ford and GM. An enterprise begins a venture in a business sector only if it anticipates a high annual sales growth rate. This does not mean that the sector necessarily has a high sales growth rate prior to the entry of the new venture, but venture analysis must indicate the possibility of a high growth rate for a superior product with unique functional features or a lower selling price or both. The combination of unique product and lower selling price is called "restructuring the market." If a high rate of sales growth is not expected, according to BCG doctrine it is foolish to begin the venture at all.

The new venture appears first as a tiny dot near the far right of the upper right quadrant of Fig. 2.2. The size of the dot represents annual sales. This quadrant is labeled

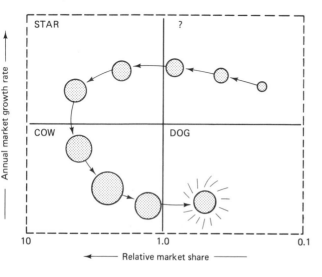

Figure 2.2 The Boston Consulting Group concept of the development of a business venture. The BCG graphing technique is unconventional in that relative market share increases to the left. Furthermore, the vertical axis is graduated in percent while the horizontal is a ratio.

with a question mark because according to the BCG doctrine the future of the venture is problematic as long as it remains in this sector. In this quadrant the cost of the venture has been relatively small and it would be possible for an enterprise to extricate itself from this market without great financial pain if anticipated sales growth does not appear.

Profits are nonexistent in this initial phase of the product life cycle because all available cash should be immediately reinvested in further growth. If the product is successful, in succeeding years the dot grows in size and moves to the left. It is vital to move leftward as rapidly as possible. For rapid growth to occur the product must have high quality and it must be presented well to the customer. If a high growth rate does not develop quickly, the project should be discontinued.

If the company has been clever in product design and all other elements of the venture, the dot grows in size and moves to the left, entering the star quadrant. Note that if a firm is not the original developer, it may be necessary to take sales from the competition to increase market share. This is as simple in practice as "belling the cat." If the total market is expanding rapidly, however, it may be possible to increase market share even as the competition also increases its sales. The implications of this latter case were first realized by BCG. This phenomena occurred in the 1960s in the U.S. motorcycle market. Table 2.1 shows U.S. motorcycle sales in the 1960s and the percentage of sales by Japanese manufacturers, principally Honda.

Table 2.1 illustrates a number of points. First, it is evident that Japanese manufacturers are not simply cutthroat competitors stealing sales by predatory pricing, as American losers want others to believe. Honda created entirely new motorcycle sales segments in this period. Honda sold motorcycles to people who had never considered buying a motorcycle before and for purposes not thought of previously, such as trail riding and off-road recreation. Second, this example indicates that even if sales are increasing, as they were for American motorcycle manufacturers in the period, a company can lose control of the marketplace and render itself vulnerable to attack. Finally, note that market share is a reliable indicator of loss of control even when gross sales figures incorrectly indicate that all is well.

TABLE 2.1   U.S. MOTORCYCLE SALES AND JAPANESE MARKET SHARE

Year	Total U.S. Cycle Sales	Sales of U.S.-built and Non-Japanese Cycles	Sales of Japanese-built Cycles	Japanese Market Share (%)
1960	45,000	42,800	2,200	5
1966	475,000 (+ 956%)	71,000 (+ 66%)	404,000 (+ 18,300%)	85
1974	1,200,000 (+ 153%)	120,000 (+ 69%)	1,080,000 (+ 169%)	90

Suppose you were sales manager of Harley-Davidson (HD) motorcycle company in 1966. As the dominant U.S. motorcycle producer would you have been pleased with the past five years' sales picture? Yes, probably so. Your sales have been increasing at a very nice 10 percent per year. Growth has not been rapid enough to require huge new capital expenditures, and all seems well. Of course, the Japanese manufacturers are also doing well, but you and the HD managers aren't worried because the Japanese aren't really competing with you. HD produces big, heavy power bikes for its loyal customers, while Japanese firms make little motor scooters and light trail bikes for kids.

Nothing that occurs in the next five years would change this position. HD growth continues at the same pace, and all seems to be well. In fact, the rapid Japanese growth prior to 1966 has now leveled off somewhat. It is a little disturbing perhaps that Japanese manufacturers have built up an excellent dealer network, have been announcing new, more powerful road machines, and seem to have plenty of money for advertising. But in only ten more years HD was bankrupt and begging President Reagan for protection so it could reorganize to meet the newly recognized threat of foreign competition.

Broadly speaking, how does a company increase sales and market share in a competitive, free market? Not only by salesmanship and advertising, as any competent marketing expert knows. A company must also offer a product with a higher value in use to the customer than the competition's. *Value in use* is defined as the cost to the customer of meeting in the best alternative way the functional need that the product is designed to serve. (See Sec. 4.2 for more details.)

We can produce a positive differential in the customer's value in use in either of two ways. First, by lowering our costs and our market price below the competitor's price. We can accomplish this by riding down the learning curve. Second, increase the product's value in use by adding unique functional features while offering our product at the same price as the competitor's product. The most devastating competition is offered by a combination of superior product and lower price. This strategy is called *market dominance*.

The 1981 case of dot-matrix printers for personal computers exemplifies such a one-two punch [2]. American manufacturers pioneered the market for low-priced, serial, impact printers and in 1979 held virtually the entire market for printers selling for less than $2500. Japanese firms entered the low end of this market and in two years gained a 50 percent market share of printers selling for less than $1000. By the end of 1981 they were expected to own a 75 percent share of this market, worth $200 million. By 1985 the market was estimated at $950 million. Epson is an example of a Japanese-produced, low-cost matrix printer. Epson's 1979 model at $750 became a price leader. By 1981 its lowest-cost model was $445, and Epson enjoyed a 30 percent share of the market. Late in 1981 IBM announced plans to offer an Epson printer with the new IBM personal computer. In November 1983 Epson reduced its standard FX-80 to $299. The rationalizations of American matrix printer makers as they went down to defeat before this attack were typical. "We really don't want that low-end business. You can't make money down there," they said, or, "We're after the professional user who wants a commercial-grade printer with speed, better quality, and more graphics."

These rationalizations simply aren't true. In fact what happened was that American dot-matrix printers lost their market share as a result of production snags, late deliveries, poor product reliability and high prices. The Epson was not only lower in price, but it also offered features not available on American machines. The machines were simply better engineered. For example, consider the printing head, which is key in the matrix printer because it takes the most wear and is the element most likely to break down. When the printing head breaks in an American machine, an expensive and time-consuming repair, possibly at the factory, was necessary. The Epson head is made of plastic and costs $25. It could be replaced by an untrained operator in five minutes. Price leadership and superior product quality made IBM choose Epson to manufacture the IBM PC dot-matrix printer.

Returning to the BCG matrix, a "star" product is characterized by high growth rate and high market share, but it does not necessarily return a profit. Earnings should continue to be returned to the venture to increase production, sales, and ultimately market share. Why the obsession with market share? Because as the learning curve indicates the largest producer has the lowest unit costs, and without the lowest costs, a company is at the mercy of the dominant (low-cost) producer. In fact, the star product is a dangerous one. Not only does it not return a profit, it can absorb great sums of money in plant expansion and other capital investment to accommodate its rapid sales growth. Should it fade too quickly, it may never return a profit and may endanger the stability of the entire organization.

After a period of rapid growth, the market matures and growth rate decreases. Now skillful maintenance of the "cash cow" is necessary. All the investment in the product while it was a question mark and then a star is wasted if a company does not reap long-term benefits now. Not only does the cash cow return profits to be shared with the owners, it also provides funds for internal financing for the growth of future stars. Proper management of cash cows can be viewed from a larger perspective than that of a single private sector organization, for instance, the Japanese global perspective.

In mid-1982 Apple Computer, Inc., faced a problem in managing a cash cow. The Apple II is a phenomenally successful personal computer, and at the beginning of 1982 it dominated the market, with the Radio Shack TRS-80 its only close competitor (see Fig. 2.3). But Apple's sales and market share began to slip as competitors finally came alive. Although the Apple II was priced at about $2000 in an operating package, other comparable machines were available that undercut this price significantly. Some machines on the market included complete software in the purchase price. Why did Apple fail to cut its selling price in the face of falling market share?

Perhaps the Apple organization needed the funds its Apple II model was generating to expand its facilities or to finance the Apple III production startup costs. Price cuts on the Apple II were rumored during the spring and summer of 1982, but Apple apparently decided to let its market share evaporate and to use its profits for internal growth. Apple's decision might have been influenced by the extraordinarily high cost of borrowing funds in this period.

In the final stage, the slow-growth, high-share cash cow product is challenged, beaten in the market by competitive products, and slips to the "dog" category. BCG doctrine calls for killing or selling the dog, but this isn't easy for corporate defenders of old faithful: "After all," they say, "this product has been a mainstay of our organization for many years. Why not stick with it and give it an extra advertising push?" The answer is that it is too late. The company has to extend the life of the product while it still enjoys

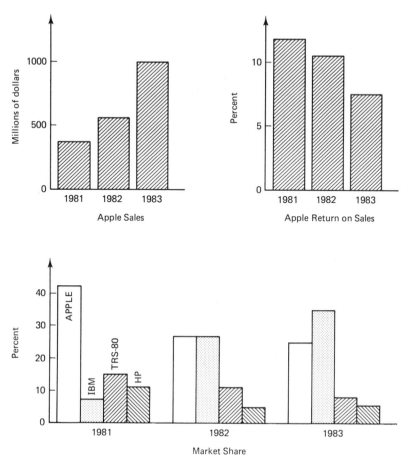

Figure 2.3   Desktop computer market share. Note that Apple lost market share even while gaining sales from 1981 to 1985. Return on sales (ROS) generally correlates better with market share than with sales. Apple conforms to this general observation in this period. Finally, in 1986 Apple made a general administrative overhaul, cutting costs and improving ROS.

a respectable market share. Extension of a cow's life explains the frequent reformulation of those ''new and improved'' washday products, for example.

Standard application of the BCG matrix is indicated in Fig. 2.2; however, reality is seldom as smooth as theory. For example, Fig. 2.4 compares Ford sales with General Motors total sales for the same period. Model T sales were remarkably volatile between 1909 and 1916. As Fig. 2.1 shows, overall Model T sales exploded in this period, but in Fig. 2.4 the emphasis is on sales growth—the rate of change of sales. Figure 2.1 gives a picture of glowing health, but Fig. 2.4 digs a little deeper to show major variations in the sales picture. As the BCG approach illustrates, by 1924 Ford was in trouble, and for the next ten years matters grew steadily worse. But Ford's trouble was not apparent without

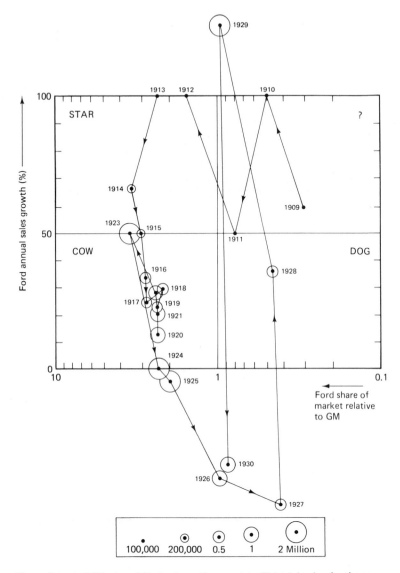

Figure 2.4 A BCG plot of Ford sales with respect to GM total sales for the same period.

close analysis. Total annual sales were high and Ford was still the market leader in this period. Precisely why do we say then that Ford was in trouble after 1924? The answer is important to managers of any high-growth company.

Ford Motor in 1924 was a company coming off 15 years of remarkable growth. Sales had exploded almost 100 times in this period. Plant capacity and employment had expanded at an astounding rate in the years before World War I. After 1915 growth continued at a rapid, yet more controlled pace, averaging 20 to 30 percent per year. Thus additional plant capacity continued to be required along with huge capital investments in production machinery. Management and labor ranks grew. These elements contributed to a management attitude sometimes known as *growth mania*.

Executives under the influence of growth mania don't have time to worry about cost controls and accounting minutiae. They focus on building factories, buying new equipment, and hiring and training personnel. Boom times are intoxicating. But in 1924 boom turned to bust. Model T sales growth disappeared and in 1925 sales showed an actual decrease. GM rapidly overtook Ford in the next few years as Ford sales continued to fall and it lost relative market share as well. In 1925 Ford began a decade of losses that mounted to a total of $26 million by 1935. Would this be tolerated in a modern, publicly held corporation? Say at Apple, for example?

But what about growth mania, has this gone as well? Probably not. Put yourself in the position of the chief executive of a large corporation whose one major product's sales have grown by a factor of almost 100 in 15 years. How would this executive adjust to the new world of declining sales? Through previous ups and downs the many critics have always been proved wrong. Why listen to them now?

The BGC matrix is useful for examining relative performance of divisions in the same organization. Figure 2.5 shows the sales performance of various divisions of GM, relative to overall corporate sales, for the period 1950 to 1984. Each chart is plotted to the same scale to facilitate comparison. Superimposed on the charts are general trend arrows. As noted in Chapter 1 the poor performance of Chevy is confirmed by the BCG matrix. Chevrolet's trend arrow moves gradually but steadily down and to the right for over 20 years. Olds and Buick, after initial wild gyrations early in the period we are examining, have settled out into steady cash cows for the corporation. Pontiac has failed for 30 years to get its act together.

Cadillac seems to hold a stable and profitable niche at the high end of the market, but it presents an inviting target for foreign luxury competitors such as Mercedes. Caddy's brand image is deteriorating, and GM doesn't seem to know what to do about it. Consult your own knowledge here. Who is the average Cadillac buyer? How old? What sex? What occupation? When asked of college students, their answers are consistent: the Cadillac buyer is male, in his middle to late fifties, a doctor or other professional person—and overall an old fogy. Do you agree? Probably so and that isn't a good position for Cadillac. Clearly, Cadillac no longer represents economic success to the middle-class American. Now that Cadillac has lost its cachet, its sales position may soon follow.

Perhaps the most interesting problem is that of Pontiac. Examine the chart. Pontiac sales have been in wild gyration for about 30 years, but Pontiac never makes progress. In the 1950s Pontiac hovered at 15 percent of total GM sales and in the late 1960s reached a peak of 20 percent. Since then, despite the weakening competition from Chevrolet and the virtual bankruptcy of Chrysler, Pontiac slipped back to its relative position of thirty years ago. And the annual variations in sales continue to be extraordinarily large. Few people can describe the typical Pontiac buyer. There is no such animal.

Clearly, Pontiac has never successfully defined its brand image. No doubt Pontiac has had good ideas in this period but it has not had the clout to be able to withstand the

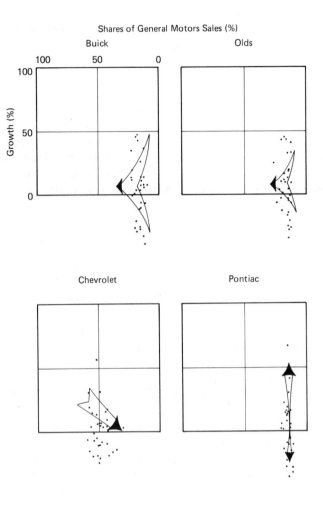

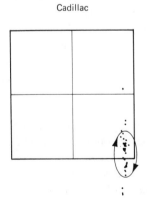

Figure 2.5 Sales growth versus sales as a percentage of total GM sales for GM divisions for the period 1950–1984.

grabs made by the Chevrolet division. Yet Chevrolet, failing to accept or be required to accept, its position in the Sloan scheme of things, has damaged not only Pontiac but itself as well. Pontiac has been described as a training ground for future general managers of Chevrolet, and if this is true it has been bad for both divisions. We are illustrating one use of the BCG chart here, but we will return to the concepts of brand image and market segmentation and treat them in more detail in Chapter 3.

2.4 THE CONGLOMERATE CONCEPT

BCG doctrine, or the gospel according to the Harvard Business School (HBS), reviewed early in Chapter 2, leads directly into the concept of the conglomerate. Here is the chain of logic. If a product is thought of as living through a growth-decay cycle, and if that product line must be eliminated when growth and market share decay as BCG requires, what follows for the organization? Obviously, the organization should have several product lines in various stages of growth. The alternative is for the entire organization to die with the dog product. This need for a varied product line is the principal motivation for the corporate diversification mania that swept the American business scene in the recent past. As HBS argues, "It is easier to buy than build" [3].

Textron is regarded as the archetype and possibly the original conglomerate.

> In 1952 Textron, Inc., a major textile manufacturer, lost $4 million on sales of $99 million. After making over 75 acquisitions and withdrawing completely from the textile industry, the company earned $44 million (after tax) on sales of $1.13 billion in 1966.

Thus opens the classic HBS case on Textron, Inc. [3]. In just these two sentences the BCG doctrine comes through. Textron rid itself completely of all business ventures in which it was technically knowledgeable but which were classified as dogs. Textron was totally out of textiles by 1966 and profitable again. But consider how radical such a concept would seem to most American managers. Do you suppose for a moment that Harley-Davidson managers could give up motorcycles?

Diversification is the driving philosophy of many major business organizations. In this sense, General Electric is a conglomerate, as is Exxon. Even U.S. Steel (now USX) and DuPont have diversified. HBS doctrine discourages its disciples from thinking of an industrial organization as a repository of special technical skills or as a venture manufacturing products in a specific area for customers loyal to its brands. A pragmatic focus on financial return is considered essential, and Textron is proof that an organization can recover from failure if it focuses in time. But a closer look at the Textron case may teach us some deeper lessons.

Royal Little, chief executive officer and a major stockholder of Textron, was determined to keep his headquarters group lean. Furthermore, he permitted no second guessing by his head staff of tactical decisions made by managers at the division level. Little's policy directly contradicted the conglomerate management style developed later at ITT under Harold Geneen. Each member of Textron's headquarters group followed a small number of Textron-owned companies and had to be ready to evaluate the annual financial plans developed by those companies. Little reserved for himself the decision to buy or sell a business, seeking staff advice, of course. Examining Little's decisions, it is possible to deduce the characteristics that make a venture a successful or unsuccessful Textron acquisition.

- Because Textron was determined not to build up a headquarters staff to manage a division and because Little was opposed to integrating divisions, it was imperative that a division have good managers in place before acquisition would be considered.
- Because Textron had to show quick financial results after purchasing a new venture, it could only consider opportunities in moderately rapidly growing fields.
- Because Textron was exclusively concerned with financial management and had no ability to supply divisions with new products from corporate research, each acquisition had to have a well-developed product line.
- Because Textron made no attempt to integrate its separate purchases, it could provide no marketing economies or savings in raw material purchases.
- Because Textron demanded high yield in the short term, it avoided high technology with its costly R&D requirements and long lead times.

It would appear that Textron sought for a rather rare opportunity. The venture under consideration for acquisition had to be well managed, at least in a technical and tactical sense, and it had to be a growing field. A desirable venture had to have good products, well-developed sources of raw materials, and good distribution and marketing programs already in place. Finally, management had to be receptive to takeover because Textron needed management after the merger. And, of course, the venture had to be available at what Little considered the right price.

Textron provided financial resources not available to smaller, less well known firms, and most important, it provided strategic management direction. Many high-technology firms are based on one good product idea from an entrepreneur with no concept of strategic business reality—the early Apple Computer, Inc., for example. Apple succeeded with the Apple II, but its next three products failed through lack of strategic marketing understanding. Textron corporate executives provided the firms it purchased with an understanding of strategic market realities.

In a long-term economic sense, however, it could be argued that Textron added little real value to its acquisitions. It simply taught existing managers how to convert long-term potential into immediate economic gain. By shifting the focus of the division managers from exclusively technical and production concerns to the bottom line and by encouraging them to forget the longer term, it may be that Textron damaged the long-term viability of these ventures. In the long term Textron would have sold out and gone on to new possibilities. Is this a stable way to conduct economic affairs? On the other hand, suppose that Textron had taken over Harley-Davidson in 1960 and enforced its financial discipline. Might Textron have saved the American motorcycle industry?

Textron's success started a trend. In the 1960s aggressive management groups began to acquire unrelated companies, and Wall Street rewarded this behavior with higher price-to-earnings ratios. "Active management of corporate assets" became the cry, and central management attention focused more on forming acquisition teams and repelling boarders than on the more fundamental problems of developing, producing, and marketing new products. There is no doubt that careful control of inventory levels and other corporate assets can lead to efficient operations and higher product quality as well as a better financial position, and perhaps that aspect of business was forgotten by many American production chiefs. Nevertheless, as with many good things, the conglomerate trend may have gone too far.

By the late 1970s reaction set in, and many chief executive officers and to acknowledge that ignorance of the fundamentals of a particular business sector did not necessarily mean they were thereby equipped with superior wisdom concerning management of a venture in the area. "Deconglomeratizing" became the new fad, with Textron and Exxon reversing direction and again leading the pack. Then in the mid-1980s Textron again switched its philosophy.

John Brooks, a well-known writer on business matters, offers a more picturesque reason for the decline of the acquisition mania in the 1970s than mere poor financial performance. In *The Game Players* he points out that when upstart Wall Street operators turned on the big banks and attempted to acquire them, the banks failed to see the humor in the situation and moved to tighten up the rules [4].

2.5 GENERAL ELECTRIC'S STRATEGIC MANAGEMENT

General Electric (GE) is a major, complex, high-technology corporation, but it does not follow the Textron pattern and should not be considered a conglomerate. Corporate management at GE does not see itself as a modern merchant bank. It is directly concerned with the management of its divisions and accepts the need for detailed technical knowledge of its activities. Yet GE's corporate diversity is greater than that of Textron and this diversity permits it to find internal funding for much of its growth. In other words, GE intends to manage its divisions, not simply observe their progress and sell them if they falter. At the same time, GE practices delegation and indeed invented a number of widely followed management practices.

The contrast being made here between diversification and a conglomerate is perhaps a little too subtle for investors to grasp. At least Jack Welch and GE believe so. In February 1989 the GE annual report for 1988 argued this point claiming its common stock was undervalued because the market did not fully appreciate the distinction [5].

General Electric was recently rated the best-managed U.S. corporation in a *Fortune* survey of corporate chief executive officers [6]. Thus its approach to strategic management should be of interest. General Electric consistently outperforms its close competitors, Westinghouse and RCA, on conventional measures of economic efficiency (see Table 2.2). Recently, GE bought out RCA.

TABLE 2.2 ECONOMIC EFFICIENCY COMPARISON OF GENERAL ELECTRIC, WESTINGHOUSE ELECTRIC, AND RCA FOR A RECENT FIVE-YEAR PERIOD

	Return on Equity		Sales Growth	
Company	Five-Year Average (%)	Rank	Five-Year Average (%)	Rank
General Electric	20.4	5	6.2	9
Westinghouse Electric	15.8	6	8.2	3
RCA	10.8	14	5.4	10
Industry median	17.7		8.7	

Industry rank refers to a *Forbes* grouping of 23 similar diversified companies.
Source: Forbes, January 14, 1985, p. 128.

GE's performance is remarkable given its huge size. Its success could only have been achieved by a vigorous, tough-minded management. In a sense GE is a conglomerate, as are others in Table 2.2, but *Forbes* excludes them from that category for the reasons just explained. *Forbes* labels this group of 23 old-line organizations that have diversified as "multi-industry companies" and labels a separate group of 45 companies "conglomerates."

Several years ago GE reorganized its $25 billion (in 1980 sales) organization into six business sectors, each headed by an executive vice-president and sector executive. The sectors are divided into groups led by a senior vice-president and group executive. There are four to six groups in each sector; on average, each group does $1 billion in sales annually. Each group also includes two to five activities headed by a vice-president responsible for $200 to $500 million in sales annually. These line VPs typically oversee two or four strategic business units (SBUs).

The SBU is a new concept that has attracted the attention of business theorists and managers of complex organizations. An SBU is defined as the smallest identifiable business unit requiring strategic planning. It consists of one or more product lines, each of which may contain a number of identifiable branded products. GE has classified its activities into 254 strategic business units. Each of these units is a large business in itself, averaging $100 million in sales. Note that the SBU is not defined around a technology nor a process. The SBU is market oriented.

The size and complexity of GE is hard to comprehend. There are over 100 VPs with line reponsibility and approximately 30 staff VPs. But GE ranked only tenth in sales in 1980, behind IBM, Ford, GM, and six oil companies. Westinghouse's 1980 sales were $8.5 billion, and RCA's were $8 billion.

GE's method of classifying SBUs for strategic planning is similar to the BCG four-element matrix. GE uses "increasing potential for growth" as one axis and "increasing centrality to GE" as the other in its diagram, shown in Fig. 2.6. Growth potential depends on both internal corporate factors and external market factors. Questions GE asks itself when evaluating growth potential include the following.

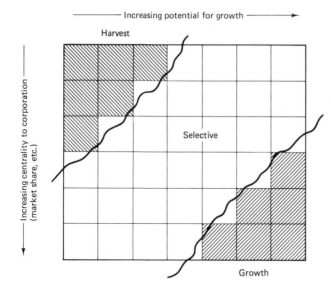

Figure 2.6 A GE-type classification of SBUs. Based on centrality and growth potential, the SBU is placed in one of three categories.

- Does our product enjoy a higher value in use than the competition's?
- Does latent demand for the product appear larger than actual demand?

Increasing centrality to the corporation is determined partly by internal and partly by external factors.

- Is the product part of a family of profitable products and necessary to support a "full line"?
- Is the product part of the "core business area" of the firm?
- Is the product supported by massive capital investment that would be difficult to liquidate?

An SBU that scores high in centrality and growth potential is classified as a growth SBU, which corresponds roughly to the question or star category in the BCG matrix. An SBU that scores low in both is classified as a "harvest" activity and is closely associated with the BCG cash cow. Mid-range SBUs are called "selective." Selective SBUs are the most complex to manage, and from them come most of the future top corporate leaders [7].

Growth and careful husbanding of corporate resources are expected from SBUs in all three categories but the relative emphasis differs. The growth SBU usually has a relatively young entrepreneur at the helm and a high proportion of the management bonus is awarded for growth in market share. The harvest SBU is generally headed by the older, more conservative manager whose bonus is more heavily weighed toward continued flow of profits. The selective SBU manager must decide which ventures to support and which to discontinue. Careful and discerning judgment along with firmness and determination are the hallmarks of the selective SBU manager.

Under Jack Welch, who became CEO in 1981, financial performance at GE has been further emphasized and nostalgia for traditional lines of business has totally disappeared. Utah International was put up for sale in 1983 and various slow-moving lines in appliances and consumer electronics have been sold off. On the other hand, GE has invested hundreds of millions of dollars in modernizing and automating its production processes. The GE home appliance factory in Lexington, Kentucky, and the diesel locomotive plant in Erie, Pennsylvania, are two examples. But GE makes capital investments in production property, plant, and equipment (PP&E) only for products that it expects will gain and then hold first or second place in market share. If GE is considered to be one of the best-managed and progressive American manufacturing corporations, its 1986 move into the service sector with the purchase of RCA and Kidder Peabody, a Wall Street brokerage firm, should give us pause.

2.6 JAPANESE NATIONAL INDUSTRIAL POLICY

It seems impossible for a small island nation to gain worldwide economic dominance. But several hundred years ago, another a small island nation gained economic and political dominance over most of the world, including what is now the United States, by virtue of superior managerial and entrepreneurial skills, a disciplined, efficient work force, and an early lead in the industrial revolution.

It is amusing to hear debates on whether "Japan, Inc.," exists. If one can accept the reality of a national consensus of banks, the government and major industries working together for the national good, then Japan, Inc., certainly exists. This consensus is developed and directed by the Ministry of International Trade and Industry (MITI), an agency of the Japanese federal government. MITI refuses export licenses to products that fail to meet its quality standards. Import licenses are refused to foreign products in protected industrial segments. Low-cost, long-term bank loans are available to entrepreneurs that venture into lines recommended by the government.

Current Japanese global economic policy developed gradually after World War II and benefited from the guidance of the MacArthur regency. It became obvious to postwar Japanese leaders that, for their nation to recover from the economic debacle brought on by its military adventurism, several conditions were necessary.

- Because Japan is devoid of raw materials, it can prosper economically only as it adds labor value to imported raw materials.
- Because the home market is small and had been stripped of its resources by the war, Japan must export or die.
- Because Asia has uncounted millions of poor, unskilled laborers living in milder climates and with easier access to local raw materials, Japan cannot compete with these underdeveloped nations by working harder. It must work smarter.

Japan started with what it had, i.e., a prewar textile industry. It began a vigorous textile export campaign in the late 1940s that contributed to Textron's loss of profits in the same period. Japanese goods competed in price, at the same time it was competing with high quality goods, excellent delivery time, and reliability. Textile exports helped Japan gather needed foreign exchange, and there is no doubt that the Korean conflict was also an economic windfall.

Japan became the staging area for the Korean Conflict and the United Nations forces built up a major presence there. Headquarters facilities, hospitals, storage depots, repair facilities all poured money into the Japanese economy and stimulated its growth.

When its high-growth textile stars fell to cash cows, Japan did not rest. It became national policy to force the textile manufacturers offshore to Hong Kong. Why? Because it was apparent that the labor value added in textiles is small and easily threatened. Textile labor is cheap and unskilled. Thus, if Japan became a high-cost producer, as it was already when compared to other Asian nations, it could not compete.

Above all, MITI saw that it should not subsidize floundering textile manufacturers to help them stay in Japan. Failing dogs could pull down the whole national economic revitalization program. It was deemed important for Japan to move up the technology curve away from dependence on textile exports. In the 1950s motorcycles and automobiles became Japan's new growth export industries. Britain faced the same problem in the 1960s with its failing motorcyle and auto industries but did not react as Japan had ten years previously in textiles. Britain made the possibly fatal error of providing massive subsidies to its dogs.

MITI encouraged Japanese manufacturers to follow its "suggestions" to move into more advanced technology by the use of government grants and manipulation of export licenses. Middle technology examples from the late 1950s and early 1960s encouraged by Japan, Inc., include motorcycles, automobiles, and steel mills. Still later, MITI encouraged these industries to move offshore and began to push home industry still further along the technology curve. It would be amusing, if it didn't hurt so, to see nations such as Britain and the United States begging Japan to export its auto factories, when MITI has decided to do that very thing in its own national interest.

Examples of high-tech industries encouraged by MITI in the 1970s and early 1980s are dry photocopiers, watches, cameras, electronics, computer printers, and high-speed facsimile devices. Future technology examples will include fifth-generation digital computers [8] and commercial jet aircraft.

The cycle continues in Asia as other nations study Japan. The same textile factories that migrated from Japan to Hong Kong in the 1950s moved to Taiwan in the 1960s. In 1972, I was at Kaohsiung in southern Taiwan, and Korean businessmen were staying at the same hotel. They were in town to buy the same textile mills to move them to Korea. In 1962, I was at Yawata Steel in Yokohama, where Japanese entrepreneurs were buying U.S.-made steel mill controls and experimenting with them on full-size blast furnaces. They were already landing superior-quality steel on the U.S. West Coast at lower prices than Kaiser Steel could produce in Los Angeles. Now, in the 1980s the Japanese are suppliers to the world of the most modern steel mills and associated controls.

Japanese labor has grown more skilled and higher priced as the nation has moved up the technology trajectory. Japan has no choice but to continue up this technology trajectory if it wishes to continue to compete in world markets. For Japan to fixate on low technology, such as textiles, or on intermediate technology, such as automobiles, would permit its Asian competition to triumph. There seem to be clear lessons for us to learn concerning the technology trajectory in the United States. U.S. industries developed the doctrines now followed by Japanese industries and we pioneered in their application. Now the student is outdistancing the teacher.

2.7 MITI MARKETING STRATEGY

Over the past several decades Japan, Inc., has developed a standard strategy for entering and developing new industrial sectors. This pattern can be seen in marketing sector after marketing sector, but it is always ignored or misinterpreted by American and European managers. From motorcycles and autos in the 1960s, through cameras and consumer electronics in the 1970s, to dry photocopiers, computer printers, and high-speed facsimile machines in the 1980s, the pattern is the same. Each time the same futile misinterpretations are heard from the competition.

"Starting at the low end" of the product price spread is the tactic perhaps most misunderstood and we will analyze it here as an example. When a Japanese entrepreneur enters a market sector with a lower-priced unit, American competitors do not usually react quickly. Trapped in internal perceptions of their own products and marketing style, U.S. managers cannot believe that Japanese low-end products can be produced at a profit.

But unlike American competitors, the Japanese product manager is thinking for the long term and is less concerned with quarter-by-quarter profit gains than with positioning the product for the future. The Japanese entrepreneur, initially financed with low-cost, long-term MITI loans, drives for global market share by anticipating the learning curve. The Japanese manager plans huge production runs in pursuit of global markets. If this means labeling the product in a half-dozen languages and making small design changes to accommodate national markets, this is acceptable, provided basic commonality is preserved.

Quality remains a high priority for the Japanese manager who intends to stay in the market as it moves upscale. Indeed the whole game is lost if the newly introduced, low-price product does not gain friends as well as sales. Customer appeal is not neglected in the low-end product. American auto manufacturers admit the superior fit and finish of the Japanese-produced car in a roundabout way through their advertising. Even in the late 1980s the United States has not caught up with Japanese quality control. In a laudatory article on the U.S. auto industry recently, *Fortune* quotes an "expert in automation in Japan and the U.S." as saying, "The best of ours is as good as the worst of theirs, and that is a tremendous achievement" [9].

Here are some other elements in a Japanese-style market assault.

- Seek only actual or potential high sales-growth market sectors. (F)
- Enter the low end first with a simple, reliable standard model. (F)
- Price the product low to gain market share rapidly. (F)
- Sacrifice immediate profits if necessary to gain and hold market share. (F)
- Concentrate on building a strong dealer network: be loyal to dealers and provide additional support through quick delivery, good parts supply, and factory service backup. (F)
- Concentrate on a high quality-to-price ratio and be alert to the potential for moving the customer upscale. (S)
- As the market matures, come in with enhanced features and higher-priced models. Move customers upscale while always giving good value for the money. (S)

Certainly these elements of the "Japanese marketing dogma" should seem familiar. Perhaps now you see better, why we started this discussion of business strategy with a review of auto history. It should be apparent that MITI learned its marketing techniques from Ford (F) and Sloan (S). If Japan, Inc., is eating our economic lunch, at least the recipe and ingredients are home-grown.

EXERCISES

1. Plot the BCG matrix for PC sales. Use Apple Computer, Inc., as the base.
2. In the 1930s the U.S. textile industry was centered in New England, where it had developed in the nineteenth century. After World War II, where did the New England textile industry migrate? Did it move by accident? Was the long-term impact on New England positive or negative? What was the long-term impact on the region where the U.S. textile industry is currently located?
3. Section 2.3 leaves Harley-Davidson at the point of bankruptcy. It did not go bankrupt. What happened and why?
4. Near the end of Sec. 2.3 Pontiac is said to have a fuzzy brand image. Pontiac management appears to have recognized this problem, according to reports published in the business press and in the period 1986–1988 Pontiac attempted to produce a focused market approach. What is this focus? Has it been successful? Did discontinuance of Fiero production in August 1988 have an impact on Pontiac's new focus?
5. Section 2.5 differentiates conglomerate and a multi-industry company. Is this a meaningful distinction? List characteristics that distinguish one type of organization from the other.
6. Suppose you were a strategic business consultant at Apple in 1981 just before the IBM PC was introduced. What would you have advised Steven P. Jobs and A. C. Markkula, Jr., concerning Apple II pricing and other strategic factors? Does it show good product planning to produce four computers (Apple II, Apple III, Lisa, and Macintosh) that exhibit mutual software incompatibility?

7. Read "Chevrolet Faces Rough Ride in Bid to Regain Sales Lead from Ford" in the *Wall Street Journal*, October 6, 1987, p. 37. Compare with Fig. 2.5 and comment on Pontiac and Chevrolet sales performance.

8. The text says that the Korean Conflict was an economic windfall for Japan. Is this true? Specifically, how did it affect Honda?

9. Harley-Davidson survives. Suppose you were asked as a consultant to H-D management for a product diversification plan for the 1990s. What would you recommend and why?

10. Update Table 2.2 with the most recent figures and comment on the changes.

REFERENCES

1. W. J. Abernathy and K. Wayne, "Limits of the Learning Curve," *Harvard Business Review*, September-October 1974, pp. 139–149.

2. "Japan's Swift Success in Printers," *Business Week*, August 31, 1981, pp. 73–74.

3. Textron, Inc., Harvard Business School Case 9-368-016 and 9-368-017 (1968), rev. 9-373-337 (1973).

4. J. Brooks, *The Games Players* (New York: New York Times Book Co., 1980), ch. 7.

5. J. F. Welch, Jr., L. A. Bossidy, and E. E. Hood, Jr., "To Our Share Owners," *General Electric Annual Report, 1988*, February 10, 1989, pp. 1, 2.

6. Ann M. Morrison, "CEOs Pick the Best CEOs," *Fortune*, May 4, 1981, pp. 133–134.

7. J. E. Gibson, *Managing Research and Development* (New York: Wiley, 1981), ch. 11.

8. *Interim Report on Study and Research on Fifth-Generation Computers* (Tokyo: Japan Information Processing Development Center, 1980).

9. "Detroit's Cars Really Are Getting Better," *Fortune*, February 7, 1987, pp. 90–98.

CHAPTER

3

Product Planning and Marketing in the Current Strategic Environment

3.1 INTRODUCTION

Corporate long-range planning, or strategic market planning, starts with an understanding of the market environment within which the organization desires to operate and the needs of the customers in this market. In choosing the specific product environment, one should take into consideration the resources of the organization that will be needed to meet customer requirements and the constraints on the organization in doing so. This market analysis indicates the specific products or services to be offered. Table 3.1 lists the steps in this planning process. The first few of these steps are generally thought of as long-range planning and the remainder as marketing. This division is artificial, however, and the process should be treated as a unit.

TABLE 3.1 STEPS IN STRATEGIC MARKETING

1. Choose general market area on the basis of corporate strengths and market opportunity.
2. Define product or service needs of potential customers in this general area.
3. Narrow the potential service area by recognizing corporate constraints such as available patent protection, labor capabilities, manufacturing capabilities, and existing distribution network.
4. Perform venture analyses to identify potential high-growth, high-profit products and services.
5. Continue with more detailed research, development, and design of candidate products.
6. Test-market and modify products on basis of test results.
7. Stay alert during early introduction to prevent surprise entry by competition.

This logical, market-oriented process is not always followed in current business practice. Quite the opposite is true, especially for technically oriented managers. Rather than beginning with what the market wants and needs, i.e., will "pull" through the distribution chain, technical managers usually start with a product they want to make and proceed backwards through the distribution chain in reverse order. This backward development, called *technology push*, is the most common approach in intermediate- and high-technology areas in which engineers are likely to have management responsibilities.

3.2 STRATEGIC MARKET PLANNING

The term long-range planning (LRP) is synonymous with the term strategic planning, or, more specifically still, strategic market planning (SMP). Although some industrial organizations do not consider strategic planning as high a priority as short-range or annual forecasting, this attitude seems to be in the process of change, and no doubt SMP will receive increasing emphasis as executives realize the implications of failing to plan.

There are a number of arguments for not planning, and the first of these is a double-edged sword. On the one hand, the executives uninterested in planning argue that business changes so fast, no plan can keep up and, on the other argue that they know the business so well that formal planning is unnecessary. Even if either or both parts of this argument seem to be true, the executive who does not plan has resigned the organization to a reactive mode. Absent planning, it can only respond to the competition and follow rather than lead. This implies not only that the competition sets prices and the direction of new model changes, and so on, but also that the organization has surrendered valuable preparation time in designing and producing new products.

A more substantive argument against SMP focuses on methodology. SMP presents difficulties because an organization establishing future directions must place priorities on how scarce resources will be allocated in the future. Naturally many groups within the organization may claim these uncommitted corporate resources, making intergroup conflict inevitable. This potential conflict can be held in check, however, provided that corporate goals are clear and the indices used to measure each group's contribution to these goals are objective and unambiguous.

A second valid argument states that detailed planning can easily lead to excessively detailed control and micromanagement by the central office. Such control can stifle the initiative and proactive attitude of middle managers.

Finally, some opponents of strategic market planning argue that individual groups, especially central office staff, can polish their own image in the planning and reporting process while not helping, even hurting, the overall progress of the organization.

This is true. It happened to a degree at ITT under Geneen, and to a greater degree at Ford Motor and the Defense department under McNamara. Yet, employees at any time and under any set of circumstances can act so as to hurt the organization and help themselves. SMP is no different in this respect from any other element of the business. One can hardly use this undisputed fact to argue against proper control of the business.

Naturally an organization should employ safeguards and checks and balances to inhibit harmful behavior throughout the company and this is true in SMP as well.

SMP methodology has grown quite elaborate in the past few years as computers have become involved, and whole texts are devoted to it. The next few sections do not become involved with such detailed ''how to do it'' matters, but instead discuss the more global aspects of corporate planning, starting with the concept of the product life cycle.

3.3 THE PRODUCT LIFE CYCLE

Section 1.4 outlines some of the factors that may inhibit growth of a product monopoly. But even if competitive factors are ignored, it seems obvious that the marketplace can absorb only a finite amount of a product. Thus the limiting or saturation effect shown in Fig. 1.3 occurs even in the absence of a successful competitor. The gestation, rapid growth, and then market saturation by a product is called the product life cycle (PLC).

In the gestation period, prior to the takeoff or explosive growth phase, the market doctrine is ill-defined or entirely absent. The entrepreneurs developing competitive products are unsure of the demographic segment to be targeted, and agreement on the features to include in the product or the price level is lacking. We will discuss this situation next in connection with the early introduction of the Apple II.

Conventional marketing doctrine for consumer products calls for a simple, basic product and a relatively high selling price in the gestation phase. Markups or margins for various elements in the distribution chain are high and the channels carefully chosen. A high percentage of marketing expenditures is devoted to generic advertising that explains the purpose of the product to the consumer. Promotional efforts, such as sales calls, are aimed at the wholesaler and the retailer rather than the consumer. The product is new and may need to be pushed through the distribution chain to the consumer. This is expensive and time-consuming and is to be avoided when possible. Specialized journals can be used to promote new high-technology products through editorial (non advertising) content.

In phase II, the rapid-growth phase, the product is refined and various models with extra features introduced. Note that Ford's Model T did not follow this element of the strategy. Selling price is reduced to increase market share and to open up new market segments. Ford performed the price reduction step beautifully with the Model T; Apple has been significantly less graceful in this regard with successive models of the Apple II. More dealers are taken on and selective channels broadened. Early selective dealers can not be expected to like this apparent abandonment, and prudent judgment is needed. Advertising expenditures should rise to keep up the push. The promotion dollar is invested in talking directly to the consumer, and market pull grows stronger and stronger.

Finally, as the market matures in phase III, the selling price is reduced to rock bottom, and dealer margins are cut to the bone. Mass distribution of the now stripped, commodity-like product should be in place, and the advertising budget is gradually reduced. This is the current PLC phase of personal computer sales.

In the PLC scenario the correlation to the BCG segmentation of ventures into question mark, star, and cow should be apparent. Although this scenario holds in the main for many marketing campaigns, there are wide variations in detail. Some products that seem to hold great promise never take off, and others that seem to have reached maturity have a second or even a third explosive growth phase after a sales plateau. Such should be true of the personal computer as hard disks, higher processing speeds, and more powerful processors are added. Often the sales plateau can be extended by reformulating the product and reintroducing it as "new" and "improved." Activity following the maturity phase is sometimes called phase IV of the PLC.

3.4 PIONEERS, EARLY ADOPTERS, MAINSTREAMERS, AND LATE ADOPTERS

Just as specific marketing tactics can be related to specific phases of the PLC of a product, so too can specific types of buyers. Market segmentation, market demographics, and psychographics are discussed later in the chapter; here we will relate buyer psychological profiles not to the "personality" of the product, nor to the economic standing of the buyer, although both of these viewpoints are also valuable, but rather to the stage in the PLC of the product itself. Table 3.2 defines four classes of buyers and relates them to the PLC.

Consider the general characteristics of each class of buyer. The *pioneer* is in love with technology. He or she, predominantly he, is a typical target for Beta-test site volunteer for new products. Pioneers enjoy putting a new product through its paces and therefore are good bug catchers. However, they are not reliable as an indicator of product acceptability among the general public. So don't use pioneers for market testing. It almost seems at times that pioneers enjoy user-unfriendly products for the challenge they provide. Products are sold techie-to-techie to pioneers and manufacturers emphasize the pioneer's insider advantage. From an employer's point of view, pioneers may waste a lot of company time on new product evaluations. If so, they are not as cost effective as they would be doing company work.

TABLE 3.2   BUYER CLASSIFICATION VERSUS PRODUCT PLC

Classification	Phase of PLC in Which Purchase Is Likely	Expected Percentage of Total Sales
Pioneers	1	1–2
Early adopters	2	5–10
Mainstreamers	Late 2–3	50–75
Late adopters	4	10–25

Early adopters are more important than pioneers to the strategic marketing plan, not only because they are more numerous but also because they have more influence with mainstreamers. The pioneer has no interest in telling other people about new products or in "showing off." The pioneer is attracted to a new product more from personal curiosity. The psychology of the early adopter is different and important for the seller to understand. It is important to the self-gratification of the early adopter to be seen and recognized as one. The early adopter is a style leader and is excellent for word-of-mouth advertising. A style leader loves to be "first on the block to own one," to talk about the new gadget just purchased, and show off to friends. The early adopter is not a techie and is not into the new product for its own merit, but rather for its image-enhancing advantages. We must interest the early adopter in our product early in phase II of the PLC, if we are to achieve rapid sales growth.

Mainstreamers are influenced by early adopters and use them as testers of new products. Mainstreamers are susceptible to endorsements and feel it is important to be in style, but they are not style leaders. Mainstreamers have little confidence in their own ability to evaluate a new product. They are not risk takers. They may fear or dislike technology. We should sell reliability and standardization to mainstreamers. Mainstreamers buy in the middle or late in phase II of the PLC and make up the whole of phase III. No early adopter wants to be caught in phase III of a PLC.

Late adopters deliberately wait until the market shakes out, to be sure of the product and to pick up bargains. The late adopter is a careful buyer and seeks the best price. The late adopter reads *Consumer Reports* and watches early adopters and mainstreamers. The late adopter is cautious, afraid of being stung, and likes money-back guarantees and free-trial offers. Late adopters make up phase IV of the PLC.

We should now be able to see how to attract each kind of purchaser. The pioneer reads the journals to keep up and attends trade shows and conferences. He seeks us out. Remember that the pioneer is not a reliable indicator of market success.

To attract the early adopter, emphasize "new," and "revolutionary, new features," as well as "be the first on your block to own one." Please don't think that we are merely talking about "washday miracles" here. I know of several electric utility chief executives who have bought multimillion dollar power plants with this attitude. Being first is important to early adopters.

To mainstreamers, sell the endorsements from early adopters and emphasize confidence in the product by mentioning the important users already in our stable. To late adopters sell price and product reliability, as well as pointing out that "everybody else has one, so get yours before it is too late."

3.5 STRATEGIC MARKET PLANNING IN GROWTH INDUSTRIES

The potential for product sales growth depends on several factors, principally product quality and price. A company may enter a market that is essentially static, but with a new, lower-cost manufacturing process and they can cut the sales price, thus taking market share from the competition. Because almost every product exhibits price elasticity, cutting the retail price not only takes business from competitors but also opens up whole new customer segments. The overall process is called restructuring the industry. In Chapter 2 we saw this happen with U.S. motorcycle sales in the 1960s.

Rather than restructure an existing business, a company may introduce a new product. Examples of new products abound: digital watches, personal computers, electronic games, video cassette recorders, instant photography, garbage compactors, automatic dishwashers, to name a few. These products are new *technologically*. However, they *replace* other products serving the same end. Thus from a market perspective they are *replacement sales*.

The first step in developing a new product is to do a market study. Venture analysis, or market study, is discussed in Chapter 4. The results of a market study are usually somewhat confusing and ambiguous, especially if the product is really new and thus difficult to describe in a few seconds during a customer survey. But if the results are sufficiently encouraging, the firm moves to the test-market phase. A small lot of the new product is produced on a pilot production line and placed with selected consumers.

Much knowledge can be gained from properly conducted test marketing. First, a company learns about the actual production costs of the product, costs difficult to anticipate without experience. Next, it finds if the consumers use the product in the way the designer intended. Entirely new uses may be discovered by watching naive but interested consumers. Further, a company also discovers the intensity of use and derives a clue to the value in use to the consumer. Finally, an approximate price elasticity curve for the product can be determined.

Test marketing involves several dangers. First, test marketing may give away part of the organization's lead over competitors. Second, there is a potential halo effect. If the tester knows your company name, she may be overly impressed and, because marketing people are by nature optimistic, they tend to overestimate the probability of success. Indeed the product may develop a halo for everyone involved with it. The Hawthorne effect can also come into play. That is, test consumers may be so flattered at the attention paid them that they are overly kind to the product under test.

Here is what F. Ross Johnson, former chief executive of RJR Nabisco, the multibillion-dollar food and cigarette firm that in 1988 ranked nineteenth among the Fortune 500, said about test marketing new products,

> You know, market researchers would run the world if they could. But I've seen plenty of products that have done well in test markets end up on their cans. Why didn't they succeed? Well, the test market in Kansas City had 2000 brand managers dusting off every shelf and knocking buck-and-a-half coupons off the price. Anyone can sell ten-dollar bills for eight-and-a-half bucks. [1]

3.6 MARKETING THE GROWTH PRODUCT

We will assume that with or without a test marketing you are about to launch the new product. A later chapter discusses the problems of the small business entrepreneur, while here we assume that you are part of a larger organization that can help with the product introduction. Venture analysis and test marketing have indicated the basic demand, and now the organization must arrange for the production and distribution of the target quantities.

While production people design the production process, marketing people must make decisions about the distribution process. Obvious arguments exist for distributing the new product using the standard company approach. However, if the product is based on a new and higher technology, your current sales force and field maintenance people are likely to resist it. National Cash Register and Univac both found it slow and difficult to move into digital computers because of resistance to "complicated" electronic products by their existing sales staff, for example. People who sell and repair mechanical devices such as adding machines and typewriters often do not take kindly to electronic marvels. Salespeople accustomed to selling luxury automobiles to affluent older people should not be expected to be successful selling compact or foreign cars to yuppies.

An organization should seriously consider setting up an entirely separate business to produce and market a new product. The values of assigning a new product to a new and separate organization are several. Managers can be invested with direct responsibility for all aspects of the business thus appealing to their entrepreneurial spirit and rewarding financially their success. Moreover, all aspects of product design, manufacture, distribution, and maintenance are addressed anew. If the president of the new company wants to use the services of the parent company, the services must be negotiated and paid for. If the new company president decides not to use the parent, the general management of the parent may wish to examine its own house.

The new organization can use the financial clout of the parent but is free to build its own production and distribution facilities as needed. The compensation plan for executives of the new organization should be light on guaranteed salary and heavy on stock options and sales bonuses. In other words, the executive who transfers to the new organization loses immediate financial benefits and security and receives in return very generous rewards for future success.

It is important that a new product be positioned properly in the market, and market segmentation is discussed later in more detail. A product can only achieve rapid sales growth if it provides a large increase in value to the consumer. A product priced too high will not sell, even if it promises significant benefits to the potential consumer. Furthermore, a high sales price erects a price umbrella under which new competitors can enter the low end of the market. The Ford Model T experience and modern Japanese competitors show that a product should initially aim for the low end of the market and then drive hard for a commanding lead.

While the top managers of U.S. industrial firms certainly understand the concept of market share and the benefits of entering the low end of the marketplace, that is not the way they usually do things. Current tax laws and the short attention span of Wall Street have produced a driving need for quick amortization of capital investment in the United States. Thus it is perhaps little wonder that managers avoid longer-term investments in new products and forgo the very path that leads to corporate longevity. They demand quick profits and set the selling price for high unit return, although this action forecloses growth and invites competition.

Demand for quick return on investment is one reason for avoiding the low end of the market and its low unit profitability. A second reason is excessively high fixed costs. Many U.S. corporations could operate more effectively with one, two, or even three levels of the management hierarchy stripped away. Japanese managers find it difficult to understand how American managers can operate so far from the factory floor. And, of course, they cannot.

American auto companies provide an example of this syndrome. Look at what Iacocca did at Chrysler. He stripped hundreds of excess workers away, especially at the managerial and professional levels, and he eliminated waste everywhere. The new Chrysler has a break-even point that is less than one-half that of the old Chrysler. Thus more than half of the "fixed" costs of the old management have been eliminated. Some critics argue that he has gone too far, yet Chrysler would have gone under if the old guard of distant managers, with their bureaucratic mentality and manicured nails, had been left in charge. The typical Japanese auto firm has seven levels of management between the production worker and the top of the corporation. General Motors has 22.

All in all, it is difficult for mature companies, with their fixed ways and high costs, to innovate and introduce high-growth potential products. But fortunately, the market mechanism and competition have a disciplining effect. Following Chrysler's lead, the entire U.S. auto industry achieved record profits in 1983 by restructuring and reducing costs to lower its break-even point by one-third [2]. In 1986, for the first time since the early 1920s, Ford Motor Company achieved higher profits than GM and repeated this feat in 1987 with less than half GM's total sales.

3.7 MARKET STRATEGY FOR MATURE INDUSTRIES

Market strategy in a mature industry seems relatively simple, involving as it does, extrapolation of the recent past, careful attention to execution and detail, and a watchful eye on competitors. As noted, the mature activity must remain in the cash cow category for as long as possible.

Obviously an enterprise in the cash cow category is not immune to assault by a new product or a more economical means of production. A more economical means of production need not imply a new process. A product produced offshore by an old process but with labor at a substantially lower wage rate is a typical threat. One can predict that this will be the form of initial threat most likely in a mature industry.

Next, competition in the form of a superior production process and or a superior product will almost certainly develop. The initial cost-cutting threat will come from Third World and developing nations bent on moving up the development curve. The product superiority threat comes from domestic competition or from Europe or Japan.

The superiority threat, in effect, presents a potential restructuring of the industry and should be dealt with as one would a new product or process. The reduced cost product threat can be quite serious. Marshall McLuhan has written of the "global village." While there is a note of fantasy in this concept, it is already a reality in commerce. A mature industry such as automobiles is no longer purely domestic. If a product enjoys a large enough market and is labor-intensive at a moderate or low skill level, its manufacturer must expect to face foreign competition. Even with patent protection, bootleg copies must be expected to cut into the world market, and the enterprise may even need to enforce its patent protection in the domestic market.

Here are some factors that can act to provide protection for a mature product from foreign invasion if utilized properly.

Capital Intensiveness.   Foreign competitors do not attempt to cross the high-entry threshold presented by a heavy capital investment unless they see a chance to restructure the marketplace. Consider what happened in the U.S. steel industry. Foreign competitors learned to produce higher-quality, lower-cost steel by investing in research and development. U.S. producers did nothing to protect themselves. R&D is noticeably absent in the U.S. steel industry. New developments such as the oxygen lance and continuous casting were created in Europe. The lesson is to protect an advantage by investing to improve productivity.

Distribution Intensiveness.   It is difficult and expensive to set up a distribution system in a nation far from one's homeland. This is said to be one of the reasons that American commercial interests find it difficult to penetrate Japanese markets. Japanese competitors do not seem to experience this problem in the United States, however.

Strong Brand Identification.   Always protect the brand. Improve quality, lower costs, and stay alert. Do not permit the product image to degrade to a commodity-like appearance.

Strong Patent Protection.   Apply for patents on all patentable developments.

Reputation for Strong Dealer and Customer Support.   If a manufacturer provides excellent repair and parts service and good dealer liaison, foreign competition finds it difficult to penetrate it's home markets.

In reviewing these elements, one is struck by the fact that these are exactly the same good-business practices that ensure the profitability of a cash cow in the face of domestic competition. There are, however, two specific factors that render a mature industry vulnerable to foreign competition.

1. *Access to Cheap, Reliable Labor*. Labor-intensive, low-technology industries are particularly vulnerable. The textile industry provides a specific example. Middle-level technology products such as iron, steel, autos, and kitchen appliances are other examples.

2. *Willingness to Accept a Longer Payback Period*. Payback period and break-even point are discussed more fully in other chapters. Briefly we can say here that American commercial enterprises expect to earn a return on investments in plant or process in a very short time, three or four years at the most.

Well-run foreign firms seem to understand that any pure strategy, such as maximizing short-run return on investment, is too simple to be correct. Risk factors must also be considered, especially the risk of losing control of basic feeder industries to foreign nations.

3.8 MARKET SEGMENTATION

The concept of market segmentation is complex and controversial in many details but is generally considered to be an important element in successful retailing market strategy. Market segmentation is the division of a product market into component parts based on buyer geographic location, buyer demography, buyer psychographics, and other factors. Buyer demography means segmentation of the market by age, sex, race, education, social conditions, and so forth. Psychographics refers to the buyer's self-image. By positioning the product properly and emphasizing the benefits a particular target population will most value, product appeal will be increased.

Marketing people believe that the purchasers of a product are "making a statement" concerning their self images by purchasing a specific brand of a particular product. By this theory, in order to maximize the appeal of a product, its "brand image" must match the consumer's self-image. This is not the sort of mental process that a technically trained person is likely to employ consciously when making a purchase, thus this concept will require careful consideration.

Failure to segment the market properly for a product will result in a blurring of the product image and a failure of the potential customer to respond. Advertising dollars are then wasted on inappropriate media, and even properly placed ads will not convey the correct overt and subliminal messages. The potential target consumers will be turned off by cues they feel are inappropriate for a particular medium, or for their self-images. Other nontargeted demographic sectors reached by the blurred ads will ignore the whole matter. Chapter 1 mentions the blurring of General Motors product images in the recent past as an unfortunate example of this phenomenon.

Chevrolet marketing represents a failure of nerve and knowledge and a major departure from the Sloan philosophy. Sloan positioned Chevy at the low end of the GM product spectrum. Originally, Chevrolet was to have features superior to the Model T and sell for a few hundred dollars more. It was to attract a slightly more affluent urban buyer than the Model T, but at the same time be the starter purchase for a first-time new car buyer. After experiencing satisfaction with its Chevrolet, the upwardly mobile American family was expected to move upscale to a Pontiac or even a Buick or an Olds. The current Japanese strategic marketing approach mirrors Sloan's philosophy.

Naturally, Chevrolet dealers resisted giving away their best customers, i.e., those capable of moving upscale, especially in tight sales years, and they pressed GM to allow Chevy to follow its customers upscale. Mediocre Chevrolet management cooperated in this blurring by arguing that the division couldn't be expected to make its return on investment (ROI) target if it continued to be cut off from the more profitable upscale sales. Because corporate management now came from the finance side, it accepted the ROI argument it understood and missed the fallacy of permitting Chevy to compete upscale with the other GM divisions.

As Chevy was permitted to market larger and more expensive models, its advertising plan had to become more complex. The division began to focus on specific models rather than Chevy as a whole. Many ads in this period didn't even mention the word *Chevrolet*. Chevy's brand image was no longer crisp. Instead of being the car for young marrieds, it tried to satisfy everyone and ended up satisfying no one. This failure in market segmentation was one cause of the GM sales disasters of the 1970s. Chevy fell from a GM market share of more than 50 percent in 1975 to less than 38 percent by 1981.

Market segmentation has been brought to its most elaborate state in the presentation of consumer goods such as body care products, snack foods, fashion clothing, and leisure time products. But marketing strategy is important for any product that experiences competition. Almost all purchases in every economic class have an important psychographical component. This is true because in American society even the vast majority of people ''below the poverty line'' live above a minimum subsistence level. In an affluent society, few purchases are made at the basic physiological survival level. Almost all purchases are inspired by higher-order needs and are therefore in some sense discretionary. Marketing in an affluent environment is a subtle game indeed.

Some economists express concern that poor people spend money on prepared products, whose retail price reflects a heavy component of expensive packaging and advertising, rather than purchasing more economical bulk generic products. For example, generic oatmeal is a warm and nutritious breakfast cereal that costs less than half the price per serving of cold packaged cereals. No better toothpaste exists than baking soda or even plain salt, but few poor people use these generics. Zealots argue that the poor do not know about the economy of generic products, but the truth seems to be that purchasing the brand name of packaged products appeals to the psychography of the buyer, regardless of economic status.

Advertising agencies perform psychographic exercises to determine the personality of the product in the market place and to position it with appropriate advertising style, price, packaging, and the outlets at which the product will be offered. The unsophisticated buyer usually does not understand that purchasing a product is based in part on the ''personality'' of the product.

Take for example the toothpaste you use. Table 3.3 gives a market segmentation of the toothpaste market. Look at the table to see if your toothpaste corresponds to your self-image. In the future, you may wish to make a point of examining print ads and TV commercials for the various brands of toothpaste mentioned in the table to see if they are designed to appeal to the defined segment.

Perhaps the most persuasive argument on the impact of advertising is its power to change the personality of a consumer product by marketing it differently. This change is possible because the brand personality is often entirely subjective to the user and exclusive of any objective product function. When Marlboro cigarettes were introduced, the brand had a red tip and was aimed at upscale women. Marlboro ads featured café society settings. The brand was a marketing failure until it was repackaged and marketed as a rugged outdoors brand for men. Note the incongruity of a London society address as the brand name for the smoke of the ''Marlboro Man.''

Technical personnel aren't usually consulted about the marketing of a consumer product, but the concept of segmentation is valid for all products. Segmentation sells cars, consumer electronics, and even heavy farm and construction equipment. If segmentation is not done properly or is ignored, the product may fail to realize its market potential.

TABLE 3.3 TOOTHPASTE MARKET SEGMENTATION

Segment	Sensory	The Sociables	The Worriers	Independents
Benefit sought	Flavor, product appearance	Brightness of teeth	Prevent decay	Price
Demographic strength	Children	Teens	Families	Men
Typical brands	Colgate Stripe	Mcleans Ultra Brite	Crest	Brand on sale
Personality traits	High self-involvement	High sociability	High hypochon-driasis	High autonomy
Life style traits	Hedonistic	Active	Conservative	Value-oriented

Source: R. I. Haley, "Benefit Segmentation: A Decision Oriented Research Tool," Table I, *Journal of Marketing, 32, July 1968, 33.*

The British motorcycle and automobile industries provide example after example of failure to understand and exploit market segmentation. In the 1950s the MG Midget was a small, economical, two-person, open sports car with excellent handling characteristics. With its leather hood strap and wire wheels it looked the part of a much more expensive, competitive sports machine. But with total disregard for its appeal to this market segment, the MG was restyled with rounded corners on its sheet metal, disk wheel covers, a homogenized instrument panel, and no hood strap. That modernization and image blurring marked a permanent downturn in the fortunes of MG. This recalls the consummate lack of marketing skill that tumbled the Land Rover from dominance in the automotive segment it pioneered.

3.9 TECHNOLOGY PUSH VERSUS MARKET PULL

Visualize if you will, a set of boxes connected in series shown in Fig. 3.1. This chain represents a simple distribution network for moving a manufactured product to the consumer. Depending on the product, the details of this chain could be altered, as is discussed in this section. For example, another level which might be called a regional distributor or a factory branch might be inserted between the manufacturer and the wholesaler, and it is possible for the wholesaler to sell at retail thus eliminating the separate retail distribution level. Even the term "manufacturer" is subject to interpretation. Who is the manufacturer of a product made by one firm for another firm whose

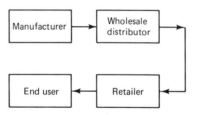

Figure 3.1   A distribution chain typical for products produced for the retail trade.

nametag appears on it? Nevertheless, Figure 3.1 represents a typical retail trade channel, although engineering goods are often moved differently.

An engineering good is one designed to meet specific, detailed performance standards, is sold by and to experts, often engineers, and which may be incorporated into a larger product or structure before being used. When a product is sold to be incorporated into a larger unit under another nameplate, it is called an original equipment manufacturer's (OEM) product. Engineering goods are distributed by independent manufacturers' representatives who carry several lines of such goods, if the manufacturer is small, or by a manufacturer's salesperson (direct rep.) if the manufacturer is large enough to maintain its own sales force.

The manufacturer in Fig. 3.1 produces a product for which he predicts a need in the retail marketplace and offers this product in bulk lots to a wholesaler at a discount from the anticipated retail price. The wholesaler purchases the product in bulk from the factory at 50 to 60 percent of the unit retail price. The wholesaler has costs associated with maintaining an inventory and delivering smaller bulk lots to the retailer. The retailer, in turn, usually buys the product from the wholesaler at 20 to 40 percent off the retail price and incurs handling and inventory costs, as well as stock obsolescence, in selling the product to the end user. Sometimes the distributor is allowed by the manufacturer to handle a number of similar products from competing manufacturers, and retailers almost always handle products from a variety of manufacturers.

Focus now on *why* the product moves through the distribution chain. Take a moment to consult your own understanding of the process. As a consumer, you develop a need or desire for a product. Perhaps you need it to perform your job, as when purchasing tools or farm machinery. Perhaps the reason is a basic physiological need for food, drink, or shelter. Or perhaps you seek to satisfy a higher-order need, the need for knowledge or peer approval, for instance. A hierarchy of human needs creates market demand, as discussed later in more detail. With this subjective need, you go to a retailer and purchase the product that maximizes your value in use. That is, you purchase the model and brand that appears most economical while meeting all of your functional needs. This definition of value in use will be sharpened in what follows.

Naturally there are times when the purchaser must compromise desires because of limited funds. The retailer also has the option of purchasing a variety of competitive products, and the ones selected are designed to maximize the retailer's value in use. The same reasoning applies to the wholesaler. This process of market pull may appear obvious, but it is not.

Many technically trained people view the process of a business venture as follows. They develop a better gadget and expect to force it through the distribution chain to the consumer by virtue of the obvious (to another technician) technological superiority of the new device. Such a technical person does not consult the market on what it wants, but instead tells the market what it should want. This process is called technology push.

Technical people sometimes advocate pushing a new technical product through the chain to the uninformed customer. The marketer on the other hand attempts to determine what the market wants, in other words, to determine the market pull. Many new product failures can be traced to the gap between these two positions. Behind almost all new products that are technically successful, but that fail commercially, can be found an engineering group and a marketing group that failed to communicate with each other. RCA has failed repeatedly in precisely this fashion. Time after time it did not bridge the gap between the design engineer and the market expert. The former CEO of RCA, Edgar H. Griffiths, admitted in an interview just after he became chief executive,

> Most of the things I've read about RCA conclude that it's poorly managed. It's a great technical company, it's a great innovator, but [then] it stumbles. The greatest thing of all is to come up with a new idea and get it to the marketplace. But then to fail because of marketing or financial reasons is absurd. And yet we've done these things.[3]

An outside observer could hardly have said it better. Yet RCA continued an inordinately risky and expensive venture in its Selectavision VideoDisc system under Griffiths, who not long afterward was forced into early retirement. The VideoDisc concept is a paradigm of the uncrossable chasm between the marketing group and the development labs at RCA, a paradigm typical of companies dominated by engineers. The VideoDisc concept was developed over a period of ten years of intense effort at the RCA Sarnoff Development Laboratory at Princeton, and it was manufactured at RCA's plant in Indianapolis, Indiana.

> For many years, the Princeton labs and the operating divisions regarded one another with deep suspicion. Jack Sauter, a superb salesman who now runs the consumer-electronics business in Indianapolis, remembers Princeton as "someplace back East where they worked on blue-sky stuff like nuclear batteries. We'd go there maybe once a year for show-and-tell." A Princeton engineer counters: "We couldn't get into the lunchroom at Indianapolis."[4]

The VideoDisc concept was fatally flawed from a consumer marketing point of view, according to some analysts. Imagine a venture analysis of the concept. One of the first things needed is to determine the value in use of the VideoDisc to the consumer. That is, with what product does the VideoDisc compete? What does it propose to replace? The VideoDisc cannot record. It is capable only of replaying factory-produced programs such as feature movies, as a phonograph record replays an audio recording. Therefore, the VideoDisc competes with movie theaters for current, first-run movies and with TV, especially Home Box Office and other cable movie offerings, for reruns. One could argue by analogy that the phonograph record has been a profitable business for more than 50 years. But when introduced, the phonograph did not have vigorous competition for the same entertainment dollar.

The most direct and vigorous competitive product with which the VideoDisc had to contend was the video cassette recorder (VCR). The VCR can record programs, including movies, from the TV to which it is attached and can be used to play rented movies at home on the TV. VCR-recorded movies could be rented for $5.00 per night in 1983 and as little as $2.00 by 1985. The VideoDisc player was offered at a discount price of about $400 in New York City in early 1982, which was close to normal dealer cost. But for $450 one could purchase a VCR at discount. RCA argued that the VCR is complex to operate and induces fear in the unsophisticated consumer. This argument seems far-fetched but even if it were true, Home Box Office on a standard TV seems a popular and economic alternative to Selectavision. In this sort of retrospective business analysis the skeptical reader is correct to suspect an author of 20/20 hindsight. At least in this one instance I can present evidence to the contrary. For a detailed venture analysis of the VCR (1977–1979), see Ref. [5]. By early fall 1982 the list price of the VideoDisc player had been reduced to $299 and RCA claimed it could make money on disk sales alone. On April 18, 1984, RCA announced that it was discontinuing production of VideoDiscs and players after suffering losses of over $500 million.

Although marketing problems do occur with sales giants such as RCA, they are even more common in smaller, newer enterprises, especially high-technology ventures. Silicon Valley is full of small, innovative firms led by brilliant and determined engineers with good ideas. Venture capital firms look for such individuals and help them with initial funding. But after the initial product startup and first marketing success, a hiatus often develops. This barrier seems often to occur at about the $50 million annual sales level. At this level the young company has about 500 employees, and sales growth may be continuing. The company has probably gone public and has perhaps introduced its second or third product, but somehow costs seem to be increasing and profits evaporating.

This situation develops from management failure to shift from an engineering-driven approach to a market-driven style. Although this is a standard syndrome, it is a coincidence perhaps that it developed more or less simultaneously with a number of minicomputer manufacturers in the late 1970s [6]. General Automation, Inc.; Computer Automation, Inc.; Microdata Corp.; and Modular Computer Systems, Inc. (Modcomp), all did well in the early 1970s as the total minicomputer market segment sales revenues increased 30 to 50 percent annually. But the members of this second-tier group were growing at less than the market rate. Industry leaders such as Digital Equipment Corp. (DEC), Hewlett-Packard (HP), and Data General grew faster than the market rate, thus not only increasing sales but also market share. As *Business Week* pointed out, "This industry growth has required management skills beyond those of the [engineering] entrepreneurs who launched many of these companies." Another comment made in the *Business Week* article about Modcomp applies to all. "The company did not go out and see what the market needed. Instead, they came up with a machine and then went out to look for a place to sell it"—a classic definition of technology push [6].

The minicomputer business is a rapidly growing industry, and opportunities are great, but marketing weaknesses coupled with a lack of adequate financial controls for inventories and accounts receivable threatens the viability of several of these second-tier manufacturers. One remedy for this sort of top management failure often resorted to by venture capital financiers of the faltering concern is to remove the engineer-founder of the firm and replace him as CEO with a financial manager from the outside.

3.10 A THEORY OF ADVERTISING

The discussion in Sec. 3.8 of why a consumer purchases a product mentions several common human needs. A. H. Maslow [7] developed a hierarchy of human needs that addresses this question. Table 3.4 lists these needs with the most primitive and basic at the bottom, ranging upward to higher-order needs at the top.

The lower-order needs seem to be common to all primates, while the higher-order needs appear exclusively human. Advertising people number among their ranks many excellent applied psychologists able to exploit human needs to create a desire to buy a particular product. Advertising people quarrel with the expression ''create a desire,'' however. Rather than create, they would prefer to say they merely bring to the surface of one's mind an existing but latent need. It is a basic tenet in the advertising credo that advertising cannot create a need but can only crystallize an existing need into the concrete form of buying impulse.

TABLE 3.4 A REPRESENTATION OF HUMAN NEEDS, BASED ON MASLOW

Need	Explanation
Transcendental	Understanding of one's place in the cosmos
Self-actualization	Realization of one's full potential
Aesthetic	The impulse toward beauty
Cognition	The need for knowledge
Esteem	Pride in self and honor from group
Group belonging	The family and the tribe
Safety and shelter	Protection from assault and the weather
Physiological	Air, water, food

*Maslow's original hierarchy does not include the transcendental category.

In advertising, a buying impulse is considered to be either latent or apparent to the consumer. If the consumer realizes the need for a product, advertising for the product becomes a notice of availability. The best example is the telephone book's Yellow Pages. No one idly leafs through the phone book for entertainment. A consumer consults the Yellow Pages when need for a product or service is apparent. The proper form of advertising under these circumstances is one that states all of the relevant information needed by the consumer to decide on a purchase. A Yellow Pages ad should contain brand names, sizes available, range of products, availability of credit, store hours, and phone number. "Creative" advertising agencies often disdain working on this sort of advertising.

If the consumer's need is latent, however, an entirely different advertising approach is required. First, the ad must be placed where a prospective but unknowing customer will be likely to see it. To accomplish this, the advertiser matches the demographics of the medium to the demographics of the prospective purchaser. Demographics refers to distribution by age, sex, geographic location, occupation, and financial condition of both the users of the medium and the target customer. One set of demographics is used to sell soap on daytime TV; another sells beer during TV football.

One does not emphasize either price or technical data in this type of ad. Rather the advertiser focuses on the benefits to be derived from its use. We can appeal to the need for esteem and belonging by implying that a purchase of the product will either admit the buyer into the company of the select or will make the buyer more physically attractive, or both. These advertising methods deal with psychographics, meaning that the marketer must have a distinct image of the kind of person who is likely to buy the product. Once the buyer realizes his latent need, further advertising takes one of several paths, depending on the market segment the marketer is approaching.

If the marketer is attempting to increase rapidly market share, ads should emphasize the attractive price. When attempting to move the buyer up to a more expensive model, advertising should emphasize exclusive features and the prestige associated with owning the advanced model, as well as increased safety and convenience to the user. (Concepts of market segmentation are discussed in Sec. 3.8.)

Advertising people tend to be bright, quick, and articulate, and they often appear emotional and uninterested in practical details—in other words, almost the direct antithesis of the technician. Thus, a technical person must be careful not to dismiss what advertising people do or how they do it. There is little these two psychological types have in common, and this can lead to disrespect if one is not careful. A good way to become acquainted with the elements of advertising is to read the best book written on it, which in the opinion of many is *Confessions of an Advertising Man* by David Ogilvy [8]. It has been published in several editions since it first appeared in 1963, including paperback, and is well worth one's time. Jerry Della Femina has written a more recent, and completely irreverent book, possibly more representative of recent advertising trends [9].

3.11 MARKETING APPLES

On January 3, 1976, Apple Computer, Inc., was founded by Steven P. Jobs, then 22 years of age, Steven Wozniak, 26, and A. C. Markkula, Jr., already a millionaire at 33 [10]. Apple is universally credited with developing the first practical personal computer, and its early financial success was remarkable. Apple sales began a rapid takeoff in 1977 with the introduction of the Apple II and have increased every year since. In 1983, after seven years of operation, sales were about $1 billion, up almost 70 percent over 1982 sales.

Yet Apple's 1983 annual report revealed a fair amount of concern for the future. Although Apple's net income and earnings per share increased in 1983, these improvements were less than half the percentage sales increase. Return on net sales, return on equity, and return on assets showed substantial decreases. Furthermore, Apple's market share continued to decline under the impact of IBM PC sales success. Figure 3.2 shows the cover sketch from Apple's 1983 annual report. This cover indicates considerable improvement over the previous year in Apple's corporate understanding of market strategy, but to some observers was still unduly optimistic.

Apple's official corporate view in 1983 expected Apple and IBM to continue to compete and share market dominance. Other domestic competitors were expected to fall by the wayside. Japanese competition was predicted to invade the PC market in 1986. How do Apple's predictions square with subsequent events in the PC marketplace?

Introduction of the Apple II in 1977 helped define the personal computer market. Still, the market initially lacked focus. Early sales were to the technical avant-garde, who found the Apple II's power, economy, and flexibility very appealing. The Apple II was a fantastic technical achievement, but it was a solution looking for a problem. Apple drifted until 1979, when the fortuitous introduction of the VisiCalc spreadsheet software

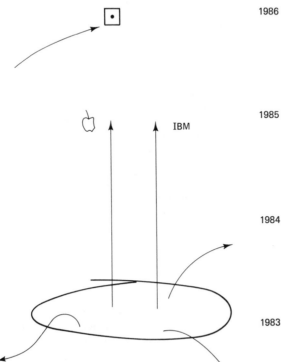

1986

1985   IBM

1984

1983

Figure 3.2 Market strategy sketch from the cover of the 1983 annual report of Apple Computer, Inc. (© Apple Computer, Inc. Used with permission.)

package designed for the Apple II, caught fire in the business market. It has been estimated that well over one-half of total Apple II sales were made in conjunction with VisiCalc. Yet Apple did not anticipate these business sales and did not move quickly to take advantage of this sales breakthrough and crucial market segment definition.

Early business adopters in lower and middle management forced Apple II and VisiCalc onto the business world, against the objections of almost the entire business establishment. Data processing center directors would not relinquish control of database management, and upper-level managers would not approve introduction of so-called computing toys into the real world of management. As a result, many early adopters bootlegged purchase of their Apples with VisiCalc, bringing the new tool in by the back door. But the smashing power of this new tool and the incredible speed with which it supplied answers to management questions overwhelmed objections to its use.

Soon thereafter the small computer market began further to segment itself. Hewlett-Packard and others came in with more expensive, more powerful 16-bit machines for the professional-technical market, and Sinclair, Atari, Commodore, and others attacked the low-price, home, and game markets. But Apple still did not take control. Apple did not seem to understand that once Big Blue saw the personal computer legitimated in the business world, it could not stand aside. Apple introduced two product failures after Apple II and only in 1987 eliminated confusion on market segmentation for the Macintosh.

Largely because it failed to understand market segmentation, the newly introduced Apple III was too expensive and was marred by software and hardware glitches. Next Apple introduced Lisa, also too expensive and not targeted for a well-defined user segment. Both Apple III and Lisa are examples of scientific hubris and technological push. Apple II, Apple III, and Lisa software are all mutually incompatible. Both Apple III and Lisa were discontinued by 1985. The first Macintosh was another technical virtuosity and still another incompatible design.

Because of these expensive new-product failures and perhaps because Apple management did not understand the power of market share, the Apple II was allowed to wander in 1981 and 1982, just as the IBM PC was introduced. Apparently some consideration was given to collapsing the Apple II price umbrella, in 1981 and 1982, at least rumors of this prospect were common in the business press at the time but the need for internal capital to fund development of the Apple III, Lisa, and Macintosh and a desire to show steadily improving earnings per share caused Apple to sacrifice market share to IBM and Tandy in this critical period.

In 1985 Apple removed Jobs from active control and turned to a marketing man from Pepsi Co, Inc., John Sculley, to become its president and CEO. Sculley understands market segmentation, and the new and improved Macintosh models introduced in 1987 are considered successful. Mac II and Mac SE are "open" so as to permit installation of supplementary hardware boards, and they have been configured to support data transfer to and from DOS-standard machines. The bit-mapped (graphics-oriented) Mac has taken advantage of the laser printer and the ability of modern software to integrate graphics and text. This has gained the Mac important leverage in the business market through the development of desk-top publishing.

EXERCISES

1. On a PLC plot label the class of product features, price, margins, distribution channels, selling and advertising approach, promotion focus, and push or pull distribution style in each PLC phase.

2. On a PLC diagram place a dot to identify your estimate of the current stage of the following electrical consumer products. Trash compactor, dishwasher, color TV, window air-conditioning unit, automatic washing machine, freezer, refrigerator, electric range, microwave oven, oscillating room fan, black-and-white TV, clothes wringer, Apple II, Macintosh, and IBM PC.

3. Section 3.4 mentions a "Beta-test site." Define and explain this term.

4. Section 3.6 suggests that a new marketing organization be set up for a new product line. Look into the case of the Acura motorcar and comment.

5. List examples of retailers that handle only one product line, contrary to Section 3.9. What is the name given such retailers?

6. See Figure 3.2. How has Apple's prediction of the PC market in 1983 stood the test of time?

7. "The reorganization [of AT&T] rips apart and rearranges an enterprise with more assets, shareholders, employees, and profits than any other in the world. This dismemberment dwarfs the 1911 split-up of Standard Oil" [11]. As a basis for class discussion, consider how the new AT&T can develop a new attitude to marketing. Did the departure of Archie McGill, president of the marketing arm, interfere with this development? See also the *Wall Street Journal* article on the trials of one of McGill's chief lieutenants, W. F. Buehler [12].

   In what respects will the institutional culture (managerial attitude) have to change at Western Electric and at Bell Labs if these two organizations are to survive in the new AT&T?

8. On May 6, 1985, Richard J. Ferris, CEO of UAL Inc., the parent of United Airlines, was featured in the cover story of *Business Week* as the executive who had done the most for his company for the least pay [13]. He was the darling of Wall Street, having increased the value of United's common stock over 2½ times—from $16⅜ per share in 1981 to $42 per share in May 1985. In May 1985 UAL was described by *Fortune* as the "airline and hotel giant" that had become "a $9 billion-a-year powerhouse in the travel industry" [14]. Ferris had led an acquisitions initiative and was described as coming "to grips with one of the airline industry's most serious problems—high labor costs." But in July 1987, *Fortune* headlined, "How Dick Ferris Blew It" [15]. Ferris had been fired, and his board of directors announced an about-face in the corporation's vision of its future. What happened here? Was Ferris wrong? Was the board wrong? Was the United pilots' union wrong? Was Wall Street wrong?

REFERENCES

1. "Ross the Boss Speaks Out," *Fortune,* July 18, 1988, p. 35.
2. "Auto Makers' Earnings Are Increasing Sharply Despite Mediocre Sales," *Wall Street Journal,* December 19, 1983, pp. 1, 19.
3. "RCA's New Vista: The Bottom Line," *Business Week,* July 4, 1977, p. 39.
4. P. Nulty, "A Peacemaker Comes to RCA," *Fortune,* May 4, 1981, p. 143.

5. J. E. Gibson, *Managing Research and Development* (New York: Wiley, 1981), ch. 8.

6. "Woes for the Second Tier in Minis," *Business Week,* September 24, 1979, pp. 116, 118.

7. A. H. Maslow, *Motivation and Personality* (New York: Harper & Row, 1954).

8. D. Ogilvy, *Confessions of an Advertising Man* (New York: Ballantine, 1971).

9. J. D. Femina, *From the Wonderful Folks Who Gave You Pearl Harbor* (New York: Simon & Schuster, 1970).

10. M. Moritz, *The Little Kingdom* (New York: Morrow, 1984).

11. *Fortune,* (special issue on the breakup of AT&T), June 27, 1983, p. 60.

12. "AT&T Manager Finds His Effort to Galvanize Sales Meets Resistance," *Wall Street Journal,* December 16, 1983, pp. 1, 20.

13. "UAL Chairman Is a Bargain," *Business Week,* May 6, 1985, p 80.

14. "United Is Changing Its Flight Pattern," *Fortune,* September 30, 1985, pp. 35–37, 44.

15. "How Dick Ferris Blew It," *Fortune,* July 6, 1987, pp. 42–46.

CHAPTER

4

Venture Analysis and the Portfolio Concept

4.1 INTRODUCTION

Venture analysis is a technique for analyzing proposed new business ventures in an objective and quantitative fashion. It attempts to establish careful estimates of the sales and potential financial return to be obtained by marketing a defined new product in a clearly delineated marketplace. The particular process we emphasize here is adapted from a method originally developed by Du Pont for use by its engineering managers, and its analytic approach should appeal to readers of this book [1].

Du Pont has found it desirable to define carefully the procedures to be followed by its managers in analyzing a proposed new venture so that the results can be compared to results obtained by other new product champions on unrelated new product ventures. The Du Pont method seems to reduce venture analysis to a quantitative, repeatable process and this is an end much to be desired. However, remember that venture analysis is an attempt to predict the future. Thus elements of luck and chance remain, despite the seeming deductive nature of the process.

Although objective analysis of individual new ventures is essential, it is not sufficient. A corporation must take an integrated and synoptic view of all lines of business (LOBs) in which it is engaged to ensure that the overall mix maximizes the return to its owners. This approach is called the *portfolio concept* and is based on the strategic principles developed in Chapters 1–3.

Perhaps the best way to illustrate the venture analysis procedure and the portfolio concept is through examples. But several important terms must be defined first.

4.2 DEFINITION OF TERMS

Total Market Potential.   Total market potential is defined as the total number of potential users for which the product is functionally adequate. However, it does not account for lack of customer knowledge of the product, lack of product availability, or, most important, the potential product's price competitiveness. Thus this number is usually very large and often relatively useless.

Market Opportunity.   This number is key. It represents the total number of customers for whom the product is functionally and economically adequate. It assumes a well-informed, objective customer and is therefore somewhat abstract, and it also implies the concept of value in use.

Value in Use.   The customer's value in use equals the total cost to the customer of satisfying a need by the best available alternative to the proposed product. As with all other definitions in this process, value in use is defined with reference to the marketplace.

Note that the concept of value in use implies that every sale is a replacement sale. The technology-oriented individual may find this concept difficult to accept. It implies that there is no such thing as a new need or a new product. Every product replaces something else according to this theory. While this may not be accurate in a narrow technological sense, it makes excellent sense from a marketing point of view.

The product must show a positive marginal value in use if it is to be pulled through the distribution chain by the end user. If a proposed venture does not show a significant margin, the product should not be marketed until estimated production and distribution costs can be reduced. This reduction may require a complete redesign of the product and its manufacturing process.

Learning Curve.   Chapter 2 shows that the cost of manufacturing a product can be reduced over time through good management practices. Thus it may be acceptable for an enterprise to anticipate this cost reduction when establishing the introductory market price of the product so as to produce a positive marginal value for the end user, even if this price falls below the estimate of the cost during the early part of the production run. This aggressive pricing strategy is called anticipatory pricing if we do it, predatory pricing if attempted by any of our national competitors, and dumping if done by our international competition.

Market Segmentation.   The total market opportunity is made up of the sum of the opportunities in all of the separate market segments. It is important to define clearly and separately, each of the various market segments for a product and to address the value in use of the customer and the way the product will be marketed in each separate segment. Proper market segmentation is often the key factor in a successful marketing effort, as discussed in Chapter 3.

Market Share.   Market share is defined as that portion of the total market opportunity that a product is expected to gain. Chapter 2 mentions the importance the Boston

Consulting Group places on market share. BCG was not the first to think deeply about market share, however. Pierre Du Pont, in creating the new Du Pont company just before and after World War I, decided that a 60 percent share of the market for a particular product was about right. He wanted to be the dominant, low-cost producer but did not intend to sacrifice profit to gain more than a 60 percent dominant position.

Pierre Du Pont argued that a 60 percent share would permit the company to cut its price and maintain full production in times of reduced demand, while allowing smaller, less efficient producers to share the reduced sales and bankruptcies among themselves. If, on the other hand, Du Pont took more than a 60 percent share in normal times, Pierre felt it might have to reduce production in slow times, thus increasing costs and possibly forcing employee layoffs, which he did not wish to do. He did not contemplate remaining in any line of business in which his company was not dominant.

Critical Incident Analysis.   The basic venture analysis is based on a surprise-free scenario. That is, the status quo and normally extrapolated trends are expected to continue. But that is really quite unlikely. Therefore, critical cost and market parameters of the venture analysis and the assumptions that underlie them should be examined to determine the effects of major changes on the new venture.

Cost of Sales.   The cost of sales (COS) or cost of goods sold (COGS) are terms used on the income statement of the firm, as explained in Chapter 5. Thus the term must be used in a consistent manner in the venture analysis. Generally, the COS shown on a firm's consolidated income statement does not break out the costs of individual products. Assigning corporate overhead, marketing, and other distribution costs to individual products is not always simple. Furthermore, the COS shown on the income statement generally refers only to those costs "under the factory roof." Marketing costs and general administration costs are usually shown separately.

It does not matter if the COS figure calculated in the venture analysis is limited to "under the factory roof" costs or if it includes marketing and general administrative costs as well, provided that the following two points are observed. First, all ventures should be compared using the same cost assumptions. Second, the product under analysis should not require marketing treatment substantially different from other products.

4.3 A VENTURE ANALYSIS OF THE HOME TRASH COMPACTOR

The home trash compactor has been viewed by some market analysts as the next big-sales kitchen appliance, although this promise remains unfulfilled. After World War II first television, then the automatic washing machines and dryers gained popularity, followed by the automatic dishwasher and the sink-mounted garbage disposal. Recently, the microwave oven was successfully introduced. What will be the next big home appliance sales success? Modest improvements in current products, such as the frost-free refrigerator and the automatic ice-cube maker, continue. But will another stand-alone major appliance find a place in the American home?

The personal computer and the video cassette recorder are already well on the way. I have given a venture analysis of the VCR elsewhere [2]. Here we consider the sales opportunity presented by the home trash compactor.

Ajax Metal Products Corporation is a diversified, light metal fabricator that manufactures desks and other metal office equipment for the OEM market. We will see more of Ajax in the coming chapters. Let us suppose that Ajax has retained you as a consultant to advise the company on the desirability of adding the home trash compactor to its other LOBs.

Total Market Potential.   The market potential of the home trash compactor is the total number of homes in America, initially neglecting the export market. There are approximately 70 million homes in the United States.

Total Market Opportunity.   The market opportunity is much smaller. First, eliminate all apartment dwellers because their trash is often removed as part of their rental service. Furthermore, renters are reluctant to invest in home improvements. Second, eliminate all home dwellers for whom trash removal is a free city service. Finally, remove all homeowners for whom the several-hundred-dollar purchase price is excessive. This reduces to asking the following question: How many homes in the United States have a family with an income of more than $25,000 annually? There are approximately 7 million such families in America. Approximately 4 million of these families live in suburban settings where trash collection is an extra-cost option. Next segment this marketplace into its major components, not forgetting sales to new home builders.

Market Segmentation.   The value in use of the compactor depends on the desires of the buyers in each market segment, so segment the market before calculating the value in use. The home compactor market includes the following three market segments:

1. Contractor purchases in bulk and at wholesale for installation in new homes
2. Owner retrofits into homes where trash collection is now an extra-cost option
3. Retrofits where trash collection is a free benefit paid from general tax revenues

Contractor Purchase in Bulk.   There were approximately 1.1 million housing starts in 1981, a down year economically. These 1.1 million homes represent the market potential in this segment. The market opportunity in this market segment might consist of housing starts above a $100,000 sales price. Approximately 100,000 such houses sold in 1981 [3].

Contractors will install trash compactors only if the compactors help them sell houses. Because the compactor is a relatively new idea it may require intensive concept advertising to produce a demand from potential home buyers. Once market pull develops, contractors seek the minimum first cost in the compactors they purchase for installation, unless they believe a brand name will create extra market pull for the house. To capture the contractor bulk market with a premium-priced, brand-name product, Ajax must invest in image building by means of widespread corporate image advertising in home magazines.

Retrofit.   Fortunately, higher-priced homes tend to be located in nonincorporated suburbs where tax-supported trash collection is the exception. Possibly one-quarter of families with an annual income of $25,000 or more live in homes worth $100,000 or more. If so, the market opportunity totals as many as 1.5 million homes. Ignore the market segment consisting of lower-priced homes located where trash collection is free.

Value in Use.   Suppose the trash compactor compresses four barrels of bulk trash into the volume of one barrel, a not unreasonable assumption because much household trash in modern America consists of discarded paper wrappings from food and other purchases. Glass and metal containers become rare as their relative cost to the packager goes up.

In some communities trash collectors charge up to $5.00 per barrel for weekly service. If four barrels of noncompacted trash represents a typical weekly load for a household, the compactor owner saves $15 per week in trash-hauling charges. To be conservative, let us assume a savings of $10 weekly, or $500 per year.

Now comes a subjective question. What payback period does the homeowner expect? One year? Two years? A payback of less than one year would be a powerful advertising claim: "Pays for Itself in Less than a Year!"

A two-year payback horizon seems about the longest in which the homeowner would be interested. Installation cost will not be small, because the kitchen counter will need to be modified, unless the design is a portable model. Fortunately, no plumbing connections are needed. Estimate a savings of $750 for the typical owner and set the sales price to the end user at this level. Ajax can now claim a payback period of less than two years, including cost of installation.

If the selling price equals the value in use, the product enjoys no marginal financial advantage over the competition. What arguments can be used with the end users to get them to pull the product through the distribution chain? Consider the following.

- Pride of ownership. The product is aimed initially at affluent households, beyond the needs realm psychologically and into the wants area. Appeal to early adopter purchasers on the basis of style leadership.
- Convenience. It is a convenience not to have to go outside to empty the trash as often or deal with many trash containers in the kitchen. Eliminates "unsightly mess" for people who are proud of their kitchens.
- Conservative payback period. If purchasers have a mental payback period of greater than a year and a half, they will realize cash savings over the lifetime of the unit. There is a positive marginal value in use for the product for these consumers and that should be emphasized in marketing efforts.

This initial rough estimate may seem rather sloppy to the analytically trained individual, but it highlights certain critical questions to be asked in a more extensive market survey, and it was completed in only a few hours. I wouldn't bother trying to pin down more precisely the assumed payback period in the next cut, because one can see that the results are only moderately sensitive to this parameter. But a marketer should establish more firmly the validity of the assumption of a $5-per-barrel trash charge. If these charges apply only to a few affluent locations, the whole analysis collapses.

Market Share.   Assume that trash compactors as a whole capture 10 percent of the housing contractor market and 5 percent of the retrofit market in any one year. The remaining market will not go to trash compactors but to discretional purchases in other areas. The reasoning is most families do not spend all of their disposable income on home appliances in a given year, no matter how advantageous the value in use.

If compactors get 10 percent of the housing starts over $100,000 (10,000 units) and 5 percent of 1.5 million possible retrofits (75,000 units), total annual sales will be 85,000 units.

If Ajax were first on the market with a compactor (as we are not), we would start with a 100 percent share. Our success would breed competition, and our share would drop. However, if a compactor already exists on the market, Ajax starts with zero share, and only if our product is superior in price and performance can it gain share to the 60 percent level. Figure 4.1 shows these estimates.

Cost of Sales.   Use "under the factory roof" costs for the COS number. The primary components of this cost estimate are as follows:

- Direct labor
- Raw materials and OEM subunits
- Manufacturing overhead

Manufacturing overhead is made up of these components:

- Production supervision costs
- Indirect labor
- Depreciation on plant and equipment

Ajax bases its manufacturing overhead on a ratio to direct labor (discussed in Chapter 6). However, as shown in Table 4.1, the percentages indicate that direct labor and manufacturing overhead are not the most critical items in the cost estimate. Instead, the OEM parts category is critical. We can see that a more detailed cost breakdown for OEM and raw materials will provide the most information. Table 4.2 gives this more detailed breakdown.

It is not surprising to find that more than 50 percent of the product cost is for materials. In some industry segments, electronics, for example, over 60 percent of the product cost may be allocated to parts and material. In this regard, Ajax is an assembler and distributor.

The difference between the variable costs and the factory selling price is called the gross margin, discussed in more detail in Chapter 5. Venture analysis works backward from the price to the end user using pro forma estimates and ratios known from experience.

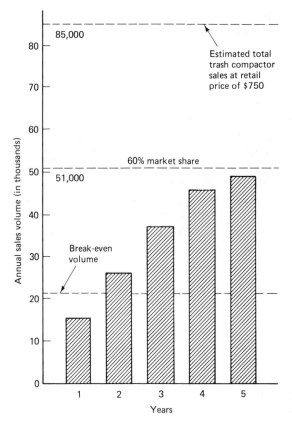

Figure 4.1 Estimated sales and market share for a home trash compactor.

It is standard for the factory price to the distributor to be about 60 percent of the end user's price. If the ratio holds, Ajax receives $450 of the $750 selling price. The U.S. Department of Commerce estimates of factory cost components in various industries. Table 4.1's breakdown is derived from Department of Commerce data for Ajax's standard industry code (SIC) category.

TABLE 4.1 MANUFACTURING COST BREAKDOWN IN THE U.S. HOME APPLIANCE INDUSTRY

	Department of Commerce Data for SIC 363 (%)	Share of $450.00 ($)
OEM and raw materials	52	234.00
Direct labor	12	54.00
Manufacturing overhead	17	76.50
Gross margin	19	85.50

Source: Annual Survey of Manufacturers, 1976 (Washington, D.C.: U.S. Department of Commerce, 1977), pp. 16–17.

To develop the parts and material breakdown in Table 4.2 requires a rough idea of how the compactor is to be constructed. The compactor consists of a rectangular trash bin that tips forward on hinges mounted on its bottom front edge to permit loading and unloading. When the bin is in the retracted position and the door latched, the electric motor may be engaged. The motor is chain-geared to sprockets mounted on the bottom of two vertical power screws. One power screw is mounted on each side of the cabinet and threads through a power nut that is in turn mounted on the compactor ram. The ram is a hollow sheet-metal block that fits down into the top of the trash bin. The power nuts cannot turn. Thus as the screws turn, the nuts must move down on the power screws, carrying with them the ram to which they are rigidly attached. Various frame elements, interlock switches, sheet-metal casing, wire, decorative trim, and hardware complete the device.

Table 4.2 estimates the gross margin to be 37 percent of the factory price, which is considerably better than the average margin in this industry sector. The next step in refining this cost estimate would be to sharpen the price estimates for the motor, ram, trash box, case, and frame. But it would be a mistake to refine the estimate before management approves the concept. If the go-ahead is given, a more careful labor estimate is prepared.

TABLE 4.2 COSTS AND GROSS MARGIN OF THE TRASH COMPACTOR VENTURE.

Cost Estimate for Trash Compactor Venture			
Our selling price to distributors			$450.00
Less direct labor (pro forma, 12%) (5.4 hr @ $10.00/hr)		$54.00	
Less manufacturing overhead (@ 140% of direct labor)		76.00	
Less OEM Materials			
Power screws @ $4.00	$8.00		
Power chain, sprockets, etc.	2.00		
1/3-hp 110-V AC motor	60.00		
Latches, switches, etc.	5.00		
Clips, wire, misc. fittings	5.00		
Ram with brackets, power nuts	20.00		
Trash box with hinges	20.00		
Case and interior frame	20.00		
		140.00	
			$270.00
Gross margin			$180.00

Suppose you were asked to present this analysis to Ajax management. Would you overemphasize uncertainty about the figures? Obviously this is a rough estimate, but it has been economical in time and talent. Management would be impatient with a request to spend more time and money to refine figures if the analyst doesn't know precisely what additional data should be collected and what is to be gained from further analysis.

4.4 CRITICAL INCIDENT ANALYSIS OF THE COMPACTOR VENTURE

The reader may wish to prepare a personal critical incident analysis of the venture in Sec. 4.3 before reading this section because this is an excellent way of learning the process. Reading someone else's words is of little help in internalizing the process.

Cost of Trash Collection.   Apparently this is the critical assumption on which the value in use calculation depends. It influences the implicit payback period used by the customer as well. In all probability, the venture analyst should recommend an informal and inexpensive market survey to establish the costs of trash collection more clearly. This cost estimate is the most critical element in predicting return on this venture. The estimate of $5.00 a barrel is a real number, but it probably holds true only in a few pockets of affluence and is out of line for the rest of the country.

Gross Margin.   Why does Ajax think that it can command a gross margin double that of its competitors in SIC category 363? If the competition could do better, wouldn't it? It really doesn't matter whether the value in use seems to be $750 if the competition is cutting the price. Let's go back and be serious. Recompute the selling price, assuming the same cost structure and a gross margin of 19 percent. By this method the retail selling price is $318.93, not $750.00, and the wholesale price is $191.36, not $450.00. This major price reduction should affect in a positive direction the sales estimate given in Fig. 4.1.

In fall 1985, a portable home trash compactor cost $239.00. So even the reanalysis above is not tight enough. Could Ajax accept a selling price of $143.40? These more realistic price estimates will produce two contrary trends. First, total compactor sales will be increased by virtue of sales-price elasticity, and second, the break-even volume increases. The first is good news for Ajax and the second is bad news. At the lowest sales price, the volume estimate in Fig. 4.1 is low by a factor of 3 to 5. Because the compactor is not a new device, we would not be surprised to find that the Department of Commerce collects sales data on it and this indeed is the case. Table 4.3 gives recent Department of Commerce sales figures for the compactor.

TABLE 4.3 SALES OF TRASH COMPACTORS (THOUSANDS OF UNITS)

1977	1979	1980	1981 (est.)	1982 (est.)
260	282	239	194	155

Source: 1983 U.S. Industrial Outlook (Washington, D.C.: U.S. Department of Commerce, 1983), pp. 42–46.

Table 4.3 confirms that price elasticity does exist and that actual compactor sales are about three times the initial estimate. Because the selling price is about one-third of the first estimate, this should not surprise us. But as a consultant, how unpleasant it would be to walk into a presentation to the management of Ajax Metal Products not knowing these data! Note that the problem with the analysis was not that the estimates were not refined in the manner of an engineering analysis. The problems with the analysis were discovered by pushing on at the same level of detail for just a few more hours. No expensive market surveys and the like were needed.

Table 4.3 indicates something else rather unpleasant. The market for compactors is not growing. This fact alone should be sufficient to kill the venture. Furthermore, information at hand is sufficient to predict that the market for compactors will not rebound in the near future. Browning-Ferris Industries, the largest and most aggressive industrial trash collection company in the country, has decided to enter the residential trash collection market. It is prepared to give the homeowner, free, a very large, wheeled trash bin that it will empty at curbside weekly for $8.00 a month. For all intents and purposes, the marginal value of a trash compactor is zero except for convenience and other nontangibles. Given these considerations, Ajax should not continue with the proposed venture at this time.

4.5 PSYCHOLOGICAL BARRIERS TO CORPORATE VENTURES

The climate is cool toward new product ventures in many large American corporations. Only when Jack Welch assumed leadership of GE in 1981 did that giant corporate leader resume product innovation on a large scale. Prior to this GE and many other business organizations felt that the short-run economic interest of the corporation was better served by acquiring small companies that had successfully introduced new products than by supporting internally developed ventures.

Even venturesome high-tech computer firms such as Digital Equipment Corporation and Data General were slow off the mark with small computer ventures, probably a result of the financial climate in the late 1970s. A company starts a venture analysis several years before the product can be introduced, and if it becomes paralyzed by high interest rates today, it will not have new products rolling out the door five years from now. The quarterly income statement may be affected by supporting new ventures, but a corporate attitude that resists product innovation faces an obsolete product line in a few years and, in the meantime, develops a banker's mentality in management that drives away managers with an entrepreneurial bent.

The best-known example of a large corporation erecting a barrier to entrepreneurial behavior is the case of the now notorious John Z. De Lorean. In 1973 De Lorean was in line to be the next chief executive officer of General Motors when he quit in frustration. Four years later he announced the organization of De Lorean Motor Car, Inc. De Lorean never abandoned the deleterious management practices common in large organizations [3]. In the end he failed in an extraordinary manner.

It was precisely the experience that De Lorean gained at General Motors headquarters *Inc.* magazine maintains, that helped render him unfit to manage an entrepreneurial organization. And precisely those same lazy and profligate corporate management techniques that tripped De Lorean led GM itself and other U.S. car makers close to the brink, contributing to their failure in the face of the Japanese motorcar challenge.

If GM had honed its top people in an entrepreneurial environment in the 1960s and 1970s, they would have been better able to handle the offshore challenges of the early 1980s. Of course GM survived, as did Ford. Even Chrysler was saved by a miracle named Iacocca and an unprecedented government bailout, but it was too close for comfort.

Suppose that in 1969, when De Lorean was promoted from his extraordinarily successful run as general manager at Pontiac to head up the faltering Chevrolet division, or that in 1972, after he started Chevy back on the right track and was promoted to corporate headquarters, he had been offered an entirely new career path: the chance to create an entirely new, independent motorcar business, using GM capital but without any of the normal corporate management controls. The more money he needed from the corporation, the more common stock he would have to surrender. The members of the management team he assembled could be cut loose from their previous corporate ties at GM with no promise of their old jobs back, or he could find new people on the outside. He could buy parts and services from the corporation, or he could go out on the open market. He could offer the new products through GM dealers who wanted to handle them, or he could set up a new dealer network.

What would De Lorean and his team gain under this totally hypothetical scenario? They would be fabulously wealthy if they succeeded, and more important, they would have an independence and power rarely possible within any corporation. GM would own a position in a profitable new venture and be able to test new markets with little risk. Futhermore, the corporation would have a valuable training ground that would instill the lean, mean management style the industry has lost. GM should offer this same opportunity to a few other fast-track managers. Some would fail and others would develop successful new market niches. Meanwhile the corporation would be training its future leaders.

The 3M corporation almost alone among major American corporations has institutionalized a positive environment toward new product development, although without forming separate corporations. IBM has taken a less extreme but similar approach recently in creating its personal computer and introducing it in record-breaking time.

Large U.S. business corporations must learn to encourage and institutionalize innovation, or they stand to lose their most creative and powerful young managers. Private venture capitalists are ready with new funding when large organizations rebuff venture ideas proposed from the inside.

4.6 AN INDUSTRY RIPE FOR RESTRUCTURING

The next few sections concentrate on an industry ready for restructuring, in fact, total reorganization. This opportunity is a result of the development of new technology and dramatic changes in the laws under which the industry operates.

On August 11, 1982, Justice Harold Greene of the U.S. Federal District Court in Washington, D.C., issued his modified final judgment (MFJ) that dismembered the American Telephone and Telegraph Corporation (AT&T) effective January 1, 1984. The judgment also provided that the surviving company could not involve itself in providing local phone service. In the same ruling Judge Greene removed the constraints on AT&T that since 1965 had prevented it from entering other activities such as the computer business.

The ruling has already had far-reaching effects and will have many others in the coming years. For example, local residential phone service will become dramatically more expensive, with rates doubling. When this occurs, it may become profitable to deliver "plain old telephone service" (POTS) to residential customers. As rate structures are modified and barriers to competition are removed, many new entries in the telecommunications business will be created. Rather than the level playing field for gentlemanly sportsmen predicted by some observers, however, the telecommunications battlefield is fast becoming a brutal melee with no holds barred and mobs of unpredictable size.

The seven regional Bell holding companies (RBHCs) and their operating companies (RBOCs), although large and dominant in their regions, are not the only providers of local phone service. In fact, there are 1400 additional independent telephone companies (telcos). Some, such as Centel and United, are almost as large as some RBOCs, and others are as small as a co-op with a few dozen subscribers. Even considering the 1400 telcos and the RBOCs, each operating in an assigned geographic area under regulated rates, does not define the new business environment, however.

A congressional staff study of the telecommunications industry characterizes its growth in the 1980s as phenomenal, far outpacing national economic, and population growth trends. In the 12-year period from 1970 to 1982 operating revenues for the telephone industry grew at a compound rate of 12 percent annually to $80 billion, while the independent (non-Bell) revenues grew at 14 percent to $14 billion. These independent telcos serve 20 percent of the U.S. population, cover 60 percent of the nation's land area, and earn 18 percent of total telephone operating revenues.

Many new and unregulated organizations have entered the telecommunications field to begin new business ventures. In the late 1960s, AT&T lost the Carterfone case, and it became legal to attach non-Bell equipment to the Bell system. Ten years later cheap, reliable digital computers revolutionized central office (CO) equipment and private branch exchange (PBX) equipment used to route calls within large organizations. It made possible automatic call waiting, call forwarding, teleconferences, and other services. Finally, the cost of high-data-rate, long-distance communication has been reduced dramatically through the use of packaged, stand-alone microwave transmission systems and communications satellites. MCI, Sprint, and others began offering alternative, lower-cost long-distance phone service, first using ground-based microwave relay links, then satellites, and most recently, fiber optic technology. A host of still newer services such as electronic mail have also been introduced.

Let's explore this newly redefined telecommunications business arena from the point of view of a small telco or individual entrepreneur thinking about a telecommunications venture. The first question to be answered is "What is the telecommunications business environment?", or "What business am I in?"

4.7 THE TELECOMMUNICATIONS BUSINESS ENVIRONMENT

Prior to deregulation, the business environment within which the typical telco operated was characterized by stability, regulated rates and conditions of service, and a complete dependence on AT&T's Long Lines Division for revenue transfer. It is not widely recognized outside the industry, just how tightly the so-called independents were in fact, tied to AT&T. Historically, by conscious regulatory decision, long-distance phone rates were set above the actual cost of delivering this service, and local phone service was provided below cost. The Federal Communication Commission (FCC) and various state regulatory commissions argued that pricing structures based on concepts such as value-of-service pricing and rate averaging better served the public than cost-based pricing. The result was low-cost residential phone service, a strong AT&T, and, perhaps not coincidentally, the best telephone service in the world.

This regulated, monopoly environment no longer exists, but it would be wrong to assign responsibility solely to Justice Greene. The historic inevitability of the arguments put forward by J. P. Morgan, who controlled the nascent AT&T, and its first general manager, Theodore Vail, to permit the original telephone monopoly in 1913 now supports deregulation. At both times the arguments were based on the available technology and the public welfare. As Sec. 4.6 demonstrates, the technology no longer requires a monopoly to run an efficient, economical telephone system.

The typical independent telco received much of its pre-1984 cash flow from funds transferred by AT&T and derived from shared long-distance billings. The carefully regulated process is called *separations and settlements*. These settlements amount to as much as 50 to 75 percent of the total income of a small independent telco. It does not take much imagination therefore, to realize how local phone rates rise as these settlements are reduced and finally eliminated. Although a few independent telcos may be content to allow corporate revenues to shrink to a fraction of their pre-1984 value, most will be driven to develop related, unregulated lines of business to replace the lost revenue stream. Fortunately, the rate structure change will be gradual. By mid-1985 local telephone rates had risen nationally only by about 15 percent. By late 1986 the increase averaged 40 percent nationwide.

The environment within which a telco formerly operated differs mightily from the environment it now faces. The old management style or "corporate culture" of the smaller telcos was patterned on the style of their former technical and financial master, the Bell System. This style could be characterized in Table 4.4.

This careful, deliberate, reactive, process mode of management is appropriate for a regulated utility, judging by the results achieved in the telephone business. However, it must be modified if the telcos, the new AT&T, and the RBOCs are to survive profitably in the new competitive, and unregulated environment.

Unique Hardware.   Telco property, plant, and equipment (PP&E) is not unique. Subscribers can buy or lease equipment elsewhere.

TABLE 4.5 POSSIBLE TELECOMMUNICATIONS-RELATED BUSINESS AREAS

Automatic redial	Fire alarm service
Burglar alarm service	Paging (beeper or message)
Bypass	Satellite home TV
Call-forwarding service	Teleconferencing
Call-waiting feature	Telemarketing
CATV (one way, multiparty)	Video classroom
Cellular radio (wireless)	Video conferencing
Computer data links	Video newspapers
Electronic mail (digital)	Videotex distribution
Electronic office	Video yellow pages
Electronic schools	Voice memo service
Emergency medical alarm	

Technical Knowledge.   Telcos do not have special technical knowledge. A telco is an operating company. AT&T set standards, and the telcos brought in outside consultants for detailed technical work. Subscribers can do likewise.

Connections to the National Network.   After deregulation, bypass, and resell, telcos have no advantageous connections to the national network.

Deregulation has created a commodity business with no special protection. But many companies deal in commodities. One brand of steel or soap or gasoline is more or less like any other, and many companies thrive in the commodity environment. The telcos need to find out how to survive in the commodity environment before they lose their subscribers.

4.9 WHAT ARE WE SELLING?

It is hard for a traditional telco manager to accept being in a commodity-like business with no monopoly protection and no unique stock in trade. But that seems to be the conclusion to which we have been forced. Let us ask then, what are some of the characteristics of successful service companies.

First and most important, the customers of successful commodity sales organizations like and trust them. Phone company subscribers trust the company. They are used to a high quality of service and reliability. They would not think of going elsewhere for POTS. However, they do not automatically think of the phone company for any of the new services listed in Table 4.5. Nevertheless, the phone company has a high level of customer confidence going for it.

Unfortunately, many subscribers do not like the phone company. Telco managers and their people know they are almost giving away local phone service, so how could subscribers think it is costly? But that's exactly what they do think. Moreover, they don't get treated nicely when they ask for service. The operators assigned to accept subscriber trouble reports at a large eastern independent telco also handle monthly billing. When busy with billing, they take their phones off the hook, and the subscriber gets a busy signal when trying to report trouble. When this problem was brought to the attention of management, a device was attached to the trouble report lines. It was discovered that only 5 percent of subscriber calls were completed. Subscribers with trouble do not like it when 19 out of 20 times the line is busy.

Management took steps to correct this condition. It lectured operators on proper procedure and installed a box that counted the number of completed calls. Soon the call completion rate rose to 95 percent. The situation did not improve, however, because operators accepted calls, said they were busy, and took the subscriber's number, promising to call back, and then never bothered to call back.

We must teach our employees that all of us are salespersons and represent the company to subscribers. We have to get the subscribers to *like* us as well as to trust us. Then we have to become proactive marketers to get them thinking about us when they want any of the extra services we can provide.

4.10 HOW TO PARSE THE OPTIONS FIELD

Table 4.5 presents some possible telco business venture options. How do telcos choose among these and other opportunities? First, they must assume a context within which to conduct the venture analysis. A venture with a heavy front-end capital requirement and a heavy continuing database maintenance cost, such as a videotex news and data system, might make excellent economic sense in a large, high-density market but makes no sense at all in the small-town, rural marketplaces of the smaller telcos.

To provide a context for an example in venture selection consider small to medium size telco called Valley Telephone (Valtelco), which has about 13,000 mostly residential subscribers and is located in a rural and small-town environment. Outside Valtelco's service region is Cassville, a city of approximately 60,000 people, and two other cities of about 30,000 each, all within a 70-mile radius. The largest city is served by a large, independent telco, and the two smaller cities are served by a smaller, independent one. Suppose that Valtelco has been awarded a cable TV franchise to serve its POTS service area and is operating a resell operation in Cassville.

To keep the regulated and nonregulated businesses separate for tax purposes and to satisfy the regulatory commission requirements regarding commingling low-cost federal loan funds granted to support rural telephone use with earnings from nonregulated business activities, in 1981 Valtelco reorganized itself into a holding corporation with three operating subsidiaries (see Fig. 4.2).

Notice that the corporate structure is new and not yet financially well balanced. The telephone company currently earns almost 98 percent of total corporate revenues.

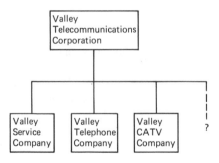

Figure 4.2 Corporate organization of Valtelco.

Nevertheless, the newer, nonregulated lines of business are vital for Valtelco's continued profitable existence.

Valtelco wants to move into nonregulated telecommunications-related businesses, but because its cash flow and borrowing power is limited, its options are limited. The following commonsense guidelines for selecting ventures apply to Valtelco and other telcos:

- Finance only one new venture at a time. Because financial resources are scarce, follow the ABC plan: *a*lways *b*e *c*areful. Be conservative and prepare detailed financial analyses of several proposed ventures before selecting one for implementation.
- Match the venture to both the financial capabilities of the organization and the portfolio of business ventures to which the telco is already committed.
- Seek a venture that permits careful, controlled entry. A venture with a vast financial threshold (relative to the telco's current size) is an invitation to disaster. Avoid ventures that must be complete before any positive cash flow is possible.
- Never go for "serial number 1." Forget about pioneering technology. Be a close follower but not a technological pioneer. Be a live and successful money maker, rather than a dead technical hero.
- Don't outdistance your people. If you staff a new venture with new people, you give away the experience advantage. Do so only with deliberate care. Employees should be trainable in the new business and be able to carry previous experience to the new job.
- Choose a venture with as short a payback period as possible and a better ROI than that obtained from reinvesting available funds in current lines of business.

Examine new ventures using techniques just presented. Be careful with the critical incident analysis. In addition, remember the BCG matrix and portfolio analysis. The POTS operation may be a temporary cash cow, but expect it to run out as settlements are reduced. Seek modest stars for new ventures. Some proposed new ventures are short-run arbitrage situations, and some are good for the long haul. Resell is an example of the former, and the electronic office appears to be of the latter class. Section 4.11 considers an appropriate strategy for combining several ventures into a well-balanced portfolio.

4.11 SHORT-TERM AND LONG-TERM VENTURES

Why do resell and bypass represent only short-term opportunities, or so-called "arbitrage" situations? First, note the technical difference between the two terms, although they are almost synonymous. The reseller is a wholesaler or a packager who buys long-distance service, WATS for example, at wholesale and retails it at a discount from the full-cost long-distance service provided by AT&T. The reseller needs only a small digital box for connecting customers to WATS lines and for keeping financial accounts of the service rendered. The reseller leases all of the equipment needed for LD message carrying from a common carrier.

The bypasser provides the same cut-rate long-distance service but also, to a greater or lesser degree, bypasses some or all of the local and long-distance facilities in favor of using the bypasser's own hardware. MCI is a bypasser for much of its traffic but a reseller for some. It is better to be a bypasser if the capital investment is justified, because the bypasser retains all fees generated from the use of owned facilities. The decision between resell and bypass turns not only on the ROI expected but also on the additional risk of committing capital equipment funds to a short-term venture.

Resell and bypass are short-run arbitrage situations rather than a long-term investment, according to the following argument. AT&T must bring down its long-distance toll fees in order successfully to compete with MCI, Sprint, Allnet, and other national bypassers. The FCC will permit this toll charge reduction in stages to preserve an orderly market and foster competition from the other common carriers (OCCs). In a few years AT&T will no longer be required to offer a discount to local bypassers and will seek that business itself.

On the other hand, the electronic office, for example, appears to be a long-term business opportunity because it represents a broad business concept rather than a specific technology. A telco entering this business sector faces hardware and software sales competition from computer retailers and systems competition from manufacturers such as IBM and Wang who sell integrated business systems as well as isolated hardware. The electronic office is generic and permits gradual entry. It will continue to develop over the next several decades and thus will support a long-term involvement. Telcos competing in this enterprise will be selling the five main components of the electronic office system, not all of which give the telco a competitive edge (see Table 4.6).

TABLE 4.6   THE FIVE MAIN COMPONENTS
OF THE ELECTRONIC OFFICE SYSTEM

- Hardware: workstations, modems, connections, and electronic files
- Software: word processors, bookkeeping systems, spreadsheets
- System installation, maintenance, and service
- System training and support
- Overall turnkey responsibility including telecommunications

4.12 VALTELCO VENTURE PORTFOLIO ANALYSIS

Relevant financial information from Valtelco's 1983 annual report and projections for possible ventures are presented in Tables 4.7 through 4.12.

Because of its small customer base, Valtelco must carefully match potential ventures with its own service characteristics. Many ventures require a large number of lines in the service area to provide reasonable market opportunity. Other ventures, such as videotex, require a major capital investment. Others, such as the video classroom, require marketing to a major population region, high capital investment, and intensive education of the potential customer base. All of these are ruled out for Valtelco.

TABLE 4.7

VALLEY TELECOMMUNICATIONS COMPANY
Consolidated Statement of Income (in thousands of dollars)

	1983	1982	1981	1980	1979
OPERATING REVENUES					
Local service	2172.5	2072.6	1937.5	1873.9	1803.3
Toll service	3675.0	2925.6	2249.9	2012.9	1818.9
Miscellaneous	365.6	297.7	269.1	221.2	182.2
CATV	123.4	80.9	17.9	—	—
Uncollectable	(31.6)	(31.4)	(17.3)	(21.0)	(9.4)
	6304.9	5345.3	4457.0	4086.9	3794.9
OPERATING EXPENSES					
Maintenance	867.3	836.3	633.1	707.1	634.3
Traffic	170.6	163.4	148.9	137.9	120.2
Commercial	347.1	236.8	209.5	168.8	150.7
General	604.6	496.2	432.6	346.0	296.6
Depreciation and amortization	1446.4	1145.5	1060.4	891.0	804.3
Other	321.5	229.5	212.5	220.7	190.3
	3757.6	3107.8	2697.1	2471.4	2196.5
OPERATING INCOME	2547.3	2237.6	1760.0	1615.5	1598.5
OTHER INCOME (net)	199.2	185.7	147.0	99.7	69.3
FIXED CHARGES*	447.5	410.1	401.2	423.5	418.2
PROFIT BEFORE TAX	2299.0	2013.2	1505.8	1291.7	1249.7
TAXES	1082.0	958.3	713.7	621.7	585.5
NET INCOME	1217.0	1054.9	792.1	670.0	664.2
EARNINGS PER SHARE†	5.20	4.51	3.37	2.86	2.79
DIVIDEND PER SHARE†	1.50	1.20	0.96	0.90	0.80

Note: Totals may not add because of rounding.

* Fixed charges are primarily interest on long-term debt.

† Per share data in dollars. Reflects 25% stock dividend in 1983.

Source: Valtelco Annual Reports, 1979–1983.

TABLE 4.8

VALLEY TELECOMMUNICATIONS COMPANY
Statistics on Lines of Business

	1983	1982	Increase	Percent Increase
ACCESS LINES	TELEPHONE			
Residential	11,140	10,797	343	3.2
Business one-party	1,136	1,093	43	3.9
Paystations	163	164	(1)	(0.6)
Business multiline	592	571	21	3.7
Total	13,031	12,625	406	3.2
TELEPHONES				
Single Line	17,163	18,827	(1,664)	(8.8)
Multiline	1,072	1,125	(53)	(4.7)
Mobile	12	15	(3)	(20.0)
Pagers	48	47	1	2.1
Paystations	163	164	(1)	(0.6)
LONG DISTANCE CALLS				
Valtelco				
Operator-handled	359,040	352,291	6,749	1.9
Direct distance dial	2,501,983	2,176,605	325,378	14.9
Total	2,861,023	2,528,896	332,127	13.1
	CABLE TELEVISION			
Customers in Eden	456	235	221	94.0
Customers in county	599	299	300	100.3
Total	1,055	534	521	97.6
	LONG-DISTANCE RESALE			
Customers in Cassville	439	—	439	—
Customers in Harrisville	186	—	186	—
Customers in Remington	154	68	86	126.5
Customers in Valley County	146	76	70	92.1
Total	925	144	781	542.4
	PLANT FACILITIES			
	Telephone	CATV		
Route miles of plant	1,375	68		
Customers per route mile	9.5	—		
Lines of dial equipment	16,040	—		
PBX and PABX	25	—		
Key systems	190	—		
Intertoll circuit to other companies	211	—		

TABLE 4.9

VALLEY TELECOMMUNICATIONS COMPANY
Income Statements of Subsidiaries
Year Ended December 31, 1983

	Valley Telephone Company ($)	Valley CATV Company ($)	Valley Service Company ($)
OPERATING REVENUE			
Local service	2,176,322	121,549	80,502
Toll service	3,636,776	—	447,736
Miscellaneous	353,965	9,348	72,674
Uncollected operating revenues	(28,504)	(346)	(2,763)
	6,138,559	130,551	598,147
Cost of sales	—	7,491	529,658
Net	6,138,559	123,060	68,489
OPERATING EXPENSES			
Maintenance	843,436	16,428	7,462
Traffic	170,570	—	—
Royalties	—	17,050	—
Commercial	231,449	2,104	96,524
General office, salaries, and expenses	529,935	9,893	69,424
Depreciation and amortization	1,395,362	44,749	4,295
Other	296,117	11,616	15,246
	3,466,869	101,840	192,951
Operating income (loss)	2,671,690	21,220	(124,462)
Other income (net)	78,501	36,507	325
Fixed charges	403,796	68,632	7,334
Profit before tax	2,346,395	(10,905)	(131,471)
Taxes (benefits)	1,069,553	(11,996)	(59,281)
Net Income	1,276,842	1,091	(72,190)

Source: Valtelco Annual Report, 1983.

A number of ventures can be added to the digital-switch technology that Valtelco has installed in its central office. Custom-calling features include call forwarding, call waiting, speed dialing, and auto redial. Other LOBs—burglar alarms, fire alarms, and emergency medical alarm services for example—also become technically feasible, but with these a telco must carefully examine its market opportunity. The potential number of subscribers might be so low that it would be difficult to develop a satisfactory ROI unless night operator labor is available at a low marginal cost.

Valtelco already operates a small CATV system in its telephone service area and a resell operation in Cassville. Compare the a priori pro forma on the CATV venture with the figures in the 1983 annual report (Tables 4.9 and 4.10), and note that this venture is slightly ahead of projections. However, equivalent comparison of the resell operation shows it well behind anticipated revenues. Perhaps it is only a coincidence, but CATV is

TABLE 4.10

	CATV Venture Pro Forma (March 30, 1980)				
Market potential	1100 establishments within the POTS service region are reachable given the minimum economic density of 12 customers per mile. Initial installation will be 45 route miles of cable and 24 establishments per mile.				
Market opportunity	(60% of potential after three years)				660
Suggested rates	One-time connection fee			$25	
	(extra charge for drops of more than 200 feet)				
	Base monthly charge			$10/mo.	
	HBO (extra charge)			$10/mo.	
Revenues (Year 3)	Town (320 signups)			$28,000	
	Rural (430 signups)			$65,000	
	Advertising			$3,600	
					$96,600
Expenses (Year 3)	$408,600 capital investment × 24% carrying charges =				$98,000
					($1,400)
Break-even point					660 signups

Estimated revenue stream	Year 1 (1981)	Year 2 (1982)	Year 3 (1983)	Year 4 (1984)	Year 5 (1985)
	(10% opp.)	(40%)	(60%)	(80%)	(100%)
	$16,000	64,000	96,600	130,000	175,000

a noncompetitive venture after the franchise is obtained, but resell is exposed to the competitive marketplace.

Valtelco's CEO is a technology leader and one of the more farsighted independent telco managers, yet even he was surprised by the aggressive entry into the Cassville resell market by two additional resellers, Sprint and Allnet. Therefore, Valtelco's market share is lower than anticipated, and the 439 customers are evenly split between business and residential.

Valtelco's CATV venture in Eden is doing well, but future growth is limited by the geographic restrictions of the franchise. Nearby towns are already served by other CATV franchisees. Valtelco can anticipate slow growth if any for this venture.

Valtelco's CEO is presented with the following five options in his portfolio analysis:

1. Keep or sell the telephone franchise.
2. Keep or sell the CATV franchise.
3. Expand, convert to bypass, or sell the resell operation.
4. Enter the electronic office LOB.
5. Enter other ventures.

TABLE 4.11

	Resell Pro Forma (September 1, 1981)	
Market opportunity	(3% of potential)	2500 subscribers
Valtelco's share	(25% of opportunity)	625 subscribers
Segmentation	50% business, 50% residential	
Revenue segment	80% business, 20% residential	
Value in use		AT&T rates
Valtelco's price		80% of AT&T rates

Object	Pro Forma Income Statement Cassville Resell*		
	Year 1 (1982), ($)	Year 2 (1983), ($)	Year 3 (1984), ($)
Revenue	270,000	740,000	1,000,000
Expenses			
Local access	40,000	95,000	120,000
WATS	200,000	460,000	600,000
Operating†	120,000	150,000	170,000
Total expenses	360,000	705,000	890,000
Net income	(90,000)	35,000	110,000

*Although resell is contemplated in this pro forma, the geographic configuration makes bypass a possibility. With resell Valtelco connects with WATS in Cassville, while for bypass it would use microwave relay link with one mountain tower giving a line of sight between Cassville and Eden. Capital expenses are estimated at $100,000 for the mountain tower and $25,000 each for the Eden and Cassville microwave terminations. Marginal bypass revenue versus resell is estimated at $0.10 additional per call.

†Digital switch capital investment for the Cassville resell venture amounts to $108,000, of which approximately 50% would be recoverable.

    Section 4.13 considers these options. Readers will find it of value to do their own analysis of the Valtelco situation using the information presented thus far, before reading further.

4.13 VALTELCO'S PLAN FOR THE FUTURE

It is unlikely that all 1400 telcos in the United States can continue to operate independently. After regulatory prohibitions disappear, significant operating economies will be available through mergers and acquisitions. Relaxation will occur as regulatory agencies come to understand that the alternatives to mergers are even less attractive. When separations and settlements evaporate, the RBOCs will become more aggressive. Thus, federal subsidies or catastrophically high local phone bills will be necessary to support the smaller telcos if they are to remain independent and economically viable.

TABLE 4.12

	Electronic Office Venture Pro Forma (March 1, 1984)		
Product description	Hardware: Computer workstations, central disk files, displays, modems, and wiring		
	Software: Word processor, mail merge, spreadsheets, and graphics		
	Maintenance and service		
	System training and support		
	System integration and telecommunications access		
Market potential	All business and professional offices in our geographic service area. (i.e., the region that we can service economically.)		
	Five workstations/business office × 2000 business offices = 10,000/stations @ $10,000/station = $100,000,000		
Market opportunity	Year 1 (1985)	Year 2 (1986)	Year 3 (1987)
	1%	5%	10%
	100 stations	500 stations	1,000 stations
Value in use			$10,000/station
Valtelco's price			$8,000/station
Valtelco's cost			40% off list in quantities of 50
			25% off list in quantities of 10

Thus, even more important for Valtelco's future than the options considered here is the need to be alert for merger opportunities with its neighbors. This process can proceed gradually, for example by initiating certain non-regulated business activities jointly with neighboring telcos. On the other hand, it could occur rapidly.

The threat of merger or acquisition is difficult for some traditional telco managers to accept. But note that in 1961 Continental Telecom began a seven-year campaign in which it purchased over 600 independent telcos, almost two a week. Recall also that in 1961 there were over 4600 independents. It might be better to merge earlier with a neighbor of similar size and retain some identity than to be swallowed whole by a giant later. But we will save a discussion on mergers and acquisitions for another time and concentrate here on growth through new independent ventures. We will consider six ventures in addition to POTS, several of which Valtelco has already entered.

Valley Telephone Company

The telephone business currently represents almost 98 percent of corporate sales and is the only business of the corporation that is a money maker. As Table 4.9 indicates, 1983 revenues were $6,138,559, and a net income of $1,276,842 represents 21 percent of revenue. The rate of return traditionally runs at about 20 percent and ROE at the same level. The telephone business currently is a cash cow, and Valtelco should not consider selling it. But separations and settlements will soon begin a decline, and Valtelco does not have many years in which to respond by creating new ventures and thus repositioning the enterprise.

Valley CATV Company

Table 4.8 reveals that the CATV venture has 1055 customers, about 495 more than initially projected. As a result, it is operating in the black a year sooner than anticipated, and prospects continue to look good. Return on sales is 1 percent, but even if it rose above 20 percent, plain cable TV could never be a big moneymaker for Valtelco unless it acquired neighboring CATV franchises at a reasonable price.

Perhaps Valtelco should hold its CATV franchise because the technology for two-way communication is developing rapidly, and it is technically feasible today to conduct school or business conferences utilizing CATV one way and the phone the other. Although Valtelco should not pioneer the commercial feasibility of these CATV enhancements, it should follow their rapid development closely, because adding value to the present CATV service will justify higher charges.

Valley Service Resell Venture

The Cassville resell venture presents a problem. Valtelco deserves high marks for early recognition of a business opportunity, but a failing grade for execution. Valtelco did not foresee the aggressive entry of two additional resellers soon after it entered the market, and this blunder had two bad results. First, it badly split the market before Valtelco had a chance to claim it, thus limiting Valtelco's ultimate market share. Second, one of Valtelco's competitors in a very clever move, offered commercial customers better prices, gaining a major business sector signup.

Valtelco's business plan projected a 50–50 mix between business and residential users, and the fee structure was set to obtain 80 percent of the total resell revenue from business users. The resell business plan assumed that a commercial customer would generate four times the revenue of a residential customer. If this ratio holds and the current revenue mix is even, Valtelco's business-residential customer mix is not 50–50 but rather is about 20–80. Because of this customer mix, revenue per customer is well below initial projections.

This represents a blunder in Valtelco's market segmentation strategy. Had the company been alert to the ratio of new customer signups after resell competition entered the Cassville market, it could have detected the trend in a month or two and taken action to restructure rates to capture a higher portion of the profitable commercial customers. But it did not do so. A television commercial Valtelco ran to introduce its resell business to Cassville showed a prosperous-looking, middle-aged actor in a residential setting telling how the viewer could save money with resell. Valtelco should have used a younger, more dynamic actor, and the setting should have been a business office to project the desired market segment subliminally.

Furthermore, Valtelco did not place sufficient weight on performing a critical incident analysis. If such an analysis had been performed, Valtelco could not have ignored the threat of competition. After all, if Valtelco saw a business opportunity, why wouldn't the competition see the same opportunity? Acting after the fact to convert a competitor's customer is much more difficult than initially signing up a new client.

We can calculate from the given data that conversion of the Cassville resell venture to bypass would have a payback period of two years. If Valtelco elects this option it could locate the microwave relay tower close to Pepperton and simultaneously invade that bypass market at a low marginal-capital cost. Risk assessment of the bypass option must focus on a careful estimate of the time horizon remaining for this line of business.

I would recommend that we avoid any further capital expenditures in connection with Cassville resell and begin an intensive six-month marketing effort focused on the top two dozen commercial accounts. Valtelco must show these accounts real cost savings and service superiority. If the situation has not turned around in nine months, Valtelco should admit failure and extricate itself from Cassville resell.

Electronic Office Venture

The sketchy outline in Table 4.12 is not sufficient to justify entry into this new line of business. Valtelco has to be careful about inadequate strategic planning. The Year 3 projection indicates a level of sales 40 percent greater than current total corporate sales and seems optimistic. If this potential truly exists, Valtelco can expect serious competitive pressures, a lesson it should have learned from the Cassville resell venture. Furthermore, because most businesses in the service area are small, is it realistic to assume five workstations per office? Reformulate the Year 2 projection to 250 stations, and consider 400 stations as steady-state annual sales starting with Year 3. However, Valtelco may not have defined the venture properly. Why not enlarge the scope of the analysis to include the education market segment?

Unit contribution will run about $3000 per station, less installation and training costs, which could be as much as $1000 per station. Fixed costs are mostly labor and therefore not fixed in the long run. Keep firmly in mind that Valtelco is not selling hardware or software, ventures that would lead to fierce competition with Radio Shack, Computerland and Entré Computer Stores. Valtelco is selling integrated communications systems and total support thereof. Servicing the aftermarket is what telco service is all about.

A knowledgeable marketing manager is vital to the success of this venture. Even more vital is a truly flexible, reliable electronic office system. Valtelco must start by converting its own corporate office procedures completely to electronic and then must run with the system it intends to sell for a quarter to make sure all the bugs are out. Before this venture goes further, Valtelco needs properly segmented and detailed monthly pro forma financial projections for the first three years of business, with gross projections for two more years. Valtelco has learned the importance of tracking variances on a monthly basis.

Cellular Radio

Cellular radio is another venture candidate to examine. Cellular ties tightly into Valtelco's current telephone business, and although the service area has a low population density, it can extend the cellular application to include towns outside its POTS area. Valtelco should take a defensive posture. If cellular costs drop and its popularity rises, competition could invade the home telephone territory. Cellular would be a good venture to consider jointly with neighboring independent telcos. Rural fixed cellular appears promising as well.

Paging Systems

Paging systems enjoy steadily increasing popularity as the technology improves and costs drop. Message paging is a low entry-cost venture that complements Valtelco's telephone service, and it should not be overlooked.

Voice Mail

The technology that supports voice mail, or voice messaging, has rapidly improved in the past few years. In 1984 a capital investment of over $500,000 was required. However, by 1987 a dozen vendors could supply personal-computer based systems in the $20,000 to $40,000 range. By 1989 the entry cost was below $10,000. Voice messaging does not require executives to change the way they conduct business, as electronic mail does. Voice mail appears to be entering the explosive growth phase in its PLC.

EXERCISES

1. The trash compactor analysis in Sec. 4.3 says, "In this regard, Ajax is an assembler and distributor." What strategic marketing and managerial implications does this statement hold for Ajax?

2. In fall 1985, Ajax Metal Products, Inc. decided not to enter the trash compactor market. But now, several years later, home construction has expanded and Ajax is reconsidering this decision. Please update the trash compactor venture analysis with current data and make your recommendation.

3. Prepare a venture analysis for a personal computer color graphics plotter. This product appeared to be an LOB ripe for exploitation in 1985. Has the situation changed?

4. In 1988 facsimile transmission of information appeared to be an LOB beginning phase 2 of the PLC. Fax machines were poised for takeoff according to many office equipment market analysts. Plot recent fax machine sales and give your estimate of their market position.

5. For several decades voice pattern recognition theory has been in development in university research laboratories and at Bell and other industrial research laboratories. In 1988 voice pattern recognition appeared poised at the beginning of phase 1 of the PLC. About 20 companies announced or introduced voice pattern recognition devices to the market. Pick a promising (specific) market segment and prepare a venture analysis for a voice pattern recognition device.

REFERENCES

1. *Dupont Guide to Venture Analysis* (Wilmington, Delaware: E. I. du Pont de Nemours Co., 1971).
2. For a venture analysis of the video cassette recorder, see J. E. Gibson, *Managing Research and Development* (New York: John Wiley, 1981), ch. 8.
3. *Inc.*, April 1983, pp. 35ff.
4. R. Baldwin, "Telco Managers Must Take the Initiative in Fighting Bypass Threat," *Telephony*, March 5, 1984.

PART

II

Managerial Finance

Part II considers the financial criteria used to judge the success of the business enterprise. The data used for such judgments are obtained from standard reports such as the balance sheet (BS) and the income statement (IS) and are processed in standard ways following the generally accepted accounting principles (GAAP).

CHAPTER

5

The Annual Report

5.1 INTRODUCTION

The annual report is the fundamental document of the business enterprise. Some business philosophers might argue for the corporate charter and some engineers might choose a basic patent owned by the company as its cornerstone, but there is almost complete agreement in the American business community that the annual report represents the company to the public. This chapter discusses a typical annual report and defines a number of ratios commonly used to test the economic stability and efficiency of an organization.

The balance sheet (BS) and income statement (IS) are universal in annual reports. In addition, all but the smallest and simplest companies feel it essential to provide an outside auditor's statement verifying that the statements in the BS and IS conform to generally accepted accounting principles (GAAP). A statement of the sources and uses of funds and a cash flow statement are valuable additions to an annual report.

The corporate annual report is prepared by the company, not by outside auditors, and it purports to reveal the financial condition of the organization to its owners. Consequently, the annual report of any large publicly held corporation contains a great deal of information. However, the information can be difficult to interpret even for people skilled in such analysis.

To be successful in a career in modern American industry, a person must have a realistic appreciation of the managerial decision process and the criteria by which business decisions are made. It seems to some observers wrong to pretend that engineers in industry make any decisions that have financial implications through a process of engineering decision making, when, in fact, they have no choice but to accept the standard accounting procedures by which the economic efficiency of the enterprise and the productivity of all managers are judged. Although we know from Chapter 1 that engineers originated almost all of the techniques used in modern management, engineering education no longer places much emphasis on this area. Indeed this neglect, it could be argued, gave primary impetus to the formation and growth of schools of business administration and the M.B.A. degree.

To illustrate how a technical person must be guided by balance sheet considerations, consider the standard problem of replacing an old but still serviceable production machine. Suppose that economic analysis indicates that the proposed shift to a new production machine is justified. It seems a matter of common sense to retain the old machine as a spare, and many production engineers would insist on doing so. Yet this option is usually ruled out because of its negative impact on the balance sheet.

Here is another problem whose solution is obvious to those familiar with the BS but which seems to contradict engineering judgment. The production engineer who "prudently" lays in a backlog of raw material and designs production lines to provide stocks of partially completed goods between machines to smooth the production process or produces for inventory to ensure prompt order filling is not prudent at all, according to modern business judgment.

From these and many more examples one could give, we are forced to conclude that engineers have no choice but to grasp the elements of managerial finance and use them to manage the industrial enterprise. Managerial finance is the universal language of business, and the balance sheet and income statement are its main tools.

5.2 THE BALANCE SHEET

The balance sheet is also called the statement of financial position and is the first of the two universal business documents discussed. Figure 5.1 shows a schematic representation of the balance sheet. The left side lists corporate assets and the right corporate equities. These two sides must be equal. By convention, liquid assets are listed at the top and more permanent assets below. The right-hand side of the BS shows corporate liabilities on top and owners' equity below. Equation (5.1) must hold if the balance sheet is truly in balance.

$$\begin{aligned} \text{Assets} &= \text{equities} \\ \text{Assets} &= \text{liabilities} + \text{owners' equity} \end{aligned} \tag{5.1}$$

One mustn't make too big a thing of the fact that the BS balances and one can't use this fact to "check" or "prove" the BS, because the way in which equality is sometimes forced by bookkeepers is to hold one entry blank until the end (so-called plug figure) and plug in a number that makes Eq. (5.1) balance.

Furthermore, one can't compare one company to another by use of the two balance sheets, because there is no single standard format and few immutable rules for construction or disclosure, despite the existence of GAAP, which tends to lend this appearance. Moreover, the manipulation of balance sheets to give the appearance of glowing health and steadily improving performance from quarter to quarter is a well-known and carefully studied (even legal) art. We will have more to say about these issues in this chapter.

```
        Assets                          Equities
*************************************************************
*                              *                              *
*  Cash          xxx   *  Payables                 xxx  *
*  Securities    xxx   *  Accruals                 xxx  *
*  Receivables   xxx   *  Taxes, accounts payable  xxx  *
*  Inventories   xxx   *     Subtotal              xxx  *
*     Subtotal   xxx   *  Debts                    xxx  *
*                      *     Subtotal              xxx  *
*                      *        Total liabilities  xxx  *
*                      ************************************
*                      *                              *
*  Investments   xxx   *  Stock                    xxx  *
*  PP&E          xxx   *  Capital surplus          xxx  *
*  Other         xxx   *  Retained earnings        xxx  *
*     Subtotal   xxx   *     Total owners' equity  xxx  *
*                      *                              *
*************************************************************

   Total assets   xxx      Total liabilities and owners' equity   xxx
```

Figure 5.1 The balance sheet in schematic form.

In Table 5.1 are listed, in order of liquidity, typical corporate assets as they might appear on the BS. The tables in Chapter 5 follow the definitions given in a well-known Merrill-Lynch pamphlet [1].

TABLE 5.1 TYPICAL ITEMS IN THE ASSET COLUMN OF A CORPORATE BALANCE SHEET

Cash Amounts in checking accounts, short-term obligations, money market funds, and other readily available instruments.

Marketable securities Holdings of interest-bearing debt, longer redemption term than cash.

Current receivables Accounts receivable represents the amount of money due from customers for goods ordered and delivered but not yet paid for. Notes receivable are moneys lent to other organizations that must be repaid.

Inventories The value at cost of raw materials, work in progress, and unsold finished goods. The LIFO and FIFO methods of costing inventory are discussed in Sec. 5.6.

Total current assets Subtotal.

Investments Funds invested for the long term in other companies, perhaps nonconsolidated subsidiaries or firms in which an organization has a majority interest.

Plant, property and equipment All capital assets, physical components of the business that will be retained from year to year. The BS may list the net figure, a current gross figure less current depreciation, or the PP&E at cost less accumulated depreciation. Depreciation is discussed in Sec. 5.5.

Other assets Optional. These entries vary from company to company.

Net fixed assets Subtotal.

Total assets

One might think that assets are good for a company and liabilities bad. But life isn't this simple. The purpose of running a business isn't to amass assets. A company is expected to use its assets to produce profits. The economic efficiency or overall productivity of an organization is often measured by the return on assets (ROA), defined as the net return (profit) divided by the assets. Thus for a given return, the smaller the asset base employed, the better.

$$\text{Return on assets (ROA)} = \frac{\text{net return after taxes}}{\text{total corporate assets}} \qquad (5.2)$$

Table 5.2 gives the liabilities that appear on a typical corporate BS.

TABLE 5.2   TYPICAL ITEMS IN THE EQUITY COLUMN OF A CORPORATE BALANCE SHEET

Accounts payable   Moneys owed by the company for goods and services bought and delivered but not yet paid for.

Notes payable   Moneys borrowed from and owed to banks in the near term.

Accrued payroll   Unpaid wages and salaries, usually a year-end accounting carry-over.

Other accrued expenses   Explained in notes appended to the financial statement.

Taxes payable

Total current liabilities   Subtotal.

Bonds, debentures   Debt instruments.

Total liabilities   Subtotal.

Preferred stock   Owners' equity. First claim, paid a fixed rate of return.

Common stock   Owners' equity. Dividends as determined by board of directors.

Capital surplus   Money paid for stock in excess of par value.

Retained earnings   Earnings retained to finance future growth.

Total owners' equity   Subtotal.

Total liabilities and owners' equity

5.3 AJAX METAL PRODUCTS

Consider the hypothetical industrial concern Ajax Metal Products Corporation. Ajax was founded in 1910 by a bright and ambitious Australian immigrant, Philip J. Freeman, and went public in the depths of the depression. It achieved modest financial success manufacturing wastebaskets for the military during World War II and has since diversified into a number of metal office product lines such as filing cabinets, desks, and chairs. Table 5.3 gives the Ajax balance sheet for a recent year.

TABLE 5.3 THE AJAX BALANCE SHEET

AJAX METAL PRODUCTS CORPORATION
Statement of Financial Position at December 31 (in thousands of dollars)

Assets	1980	1979
Current assets		
Cash	1,350	900
Marketable securities, at cost; market value: 1980, $2240	2,550	1,380
1979, $1840		
Accounts receivable		
Less allowance for bad debt:		
1980, $300		
1979, $285	6,000	5,700
Inventories	8,100	9,000
Total current assets	18,000	16,980
Fixed assets		
(property, plant, and equipment)		
Land	1,350	1,350
Buildings	11,400	10,800
Machinery	2,850	2,550
Equipment	300	285
	15,900	14,985
Less accumulated depreciation	5,400	4,500
Net fixed assets	10,500	10,485
Prepayments and deferred charges	300	270
Intangibles		
(Goodwill, patents, trademarks)	300	300
Total assets	29,100	28,035

Liabilities	1980	1979
Current liabilities		
Accounts payable	3,000	2,820
Notes payable	2,550	3,000
Accrued expenses payable	990	900
Federal income taxes payable	960	870
Total current liabilities	7,500	7,590
Long-term liabilities		
First mortgage bonds		
(5% interest, due 1985)	8,100	8,100
Total liabilities	15,600	15,690

TABLE 5.3   *(Continued)*

Stockholders' equity		
Capital stock		
Preferred stock	1,800	1,800
(5% cumulative, $300 par value each; authorized, issued, and outstanding, 6000 shares)		
Common stock	4,500	4,500
($15 par value each; authorized, issued, and outstanding, 300,000 shares)		
Capital surplus	2,100	2,100
Accumulated retained earnings	5,100	3,945
Total stockholders' equity	13,500	12,345
Total liabilities and stockholders' equity	29,100	28,035

The balance sheet presents a financial picture of the organization at one point in time, usually the close of the business year. It is difficult to gain a complete picture of the company from this single report, but some indications can be gleaned. Various ratios and standardized indices have been developed that aid in this evaluation. Many common ratios require one number from the balance sheet and the other from the income statement. These ratios are examined in Sec. 5.7. However, several indicators use data from the BS alone.

5.4 BALANCE SHEET RATIOS AND INDICATORS

Working Capital

Net working capital, net working assets, and net current assets are terms with identical meaning. Each equals the difference between the total current assets of a firm and the total current liabilities. Fortunately, the Ajax BS explicitly indicates these two factors. Some balance sheets fail to give these subtotals. Ajax's working capital can be computed directly.

Ajax current assets	$18,000,000
Less current liabilities	7,500,000
Working capital, 1980	$10,500,000

Working capital represents funds currently available to operate the business. Raw material purchases, wages, and other expenses can be met using working capital. Ajax's working capital has increased in the past year. But is $10 million high or low? The answer depends on the amount of business Ajax does annually. One measure of the adequacy of the working capital of the firm is the current ratio.

Current Ratio

The ratio of current assets to current liabilities is defined as the current ratio or liquid ratio.

$$\text{Ajax current ratio} = \frac{\$18,000,000}{\$7,500,000} = 2.4$$

Compare Ajax with some other well-known manufacturing firms to determine its relative soundness. A manufacturing company that maintains a large inventory and sells on credit requires a larger current ratio than one that does not. Table 5.4 gives the current ratio for several manufacturing firms. In addition, Table 5.4 shows inventory as a percent of the current ratio and accounts receivable (AR) as a percent of the same ratio.

General Electric is generally considered a firmly managed, large, multi–product line manufacturer. It took the lead in the late 1970s and early 1980s by aggressively managing in an inflationary economy. GE's low current ratio probably means that it is supporting its level of business with a low asset base, which is good. Note the low proportion of current assets represented by inventory at GE, another indication of aggressive asset control.

Reynolds Metals is a major aluminum manufacturer, and its relatively higher inventory probably represents raw material stocks and ore in the refinement process.

TABLE 5.4 CURRENT RATIO FOR SEVERAL MANUFACTURING FIRMS

1980	Working Capital (millions of $)	Current Ratio	Inventory as a Percentage of Current Ratio	AR as a Percentage of Current Ratio
General Electric	2,291.0	1.3	33	43
Reynolds Metals	576.7	1.9	54	40
Tandem Computers	61.2	4.0	25	50
Ajax Metal Products	10.5	2.4	45	33

Tandem is a smaller, rapidly growing computer manufacturer. It doubled current assets, AR, and inventory in 1980, yet it shows remarkable liquidity. Its cash position almost tripled in 1980. Tandem had not yet paid a dividend. Perhaps it was preparing to do so, or perhaps it needed such an extraordinary proportion of its current assets in cash to finance its rapid growth. Ajax lies in the middle range. It appears to have adequate working capital to finance reasonable growth, but if it is not growing, it may not be managing its assets with sufficient vigor. Additional information is required before completing an evaluation of Ajax.

Quick Assets

Quick assets are assets available for sudden emergencies. They consist of cash and accounts receivable, in other words, current assets less inventory. Net quick assets are quick assets less total current liabilities. Finally, the quick asset ratio, the "acid test," is quick assets divided by current liabilities. Obviously, quick asset tests are more rigorous than the current asset numbers just considered.

$$
\begin{array}{lr}
\text{Ajax current assets} & \$18,000,000 \\
\text{Less inventories} & \underline{8,100,000} \\
\text{Ajax quick assets} & \$\ 9,900,000 \\
\text{Less current liabilities} & \underline{7,500,000} \\
\text{Ajax net quick assets} & \$\ 2,400,000
\end{array}
$$

$$
\text{Ajax quick asset ratio} = \frac{\$9,900,000}{\$7,500,000} = 1.32
$$

Table 5.5 gives the 1980 quick asset ratio for Ajax and the other manufacturing firms examined.

TABLE 5.5   THE QUICK RATIO OF SEVERAL MANUFACTURERS IN 1980

1980	Quick Assets (millions of $)	Net Quick Assets (millions of $)	Quick Asset Ratio
General Electric	6,540	(1,052)	0.86
Reynolds Metals	567.7	(100.3)	0.85
Tandem Computers	60.8	40.3	2.97
Ajax Metal Products	9.9	2.4	1.32

Note: Parentheses indicate negative numbers.

One of the assumptions of GAAP is the "going concern." It is presumed the firm is in business to stay and there is no need to worry about liquidating it. Obviously this is an essential assumption for both GE and Reynolds. Neither has sufficient quick assets to meet immediate liabilities without auctioning off fixed assets. This situation would not have been considered prudent management practice in the 1960s, but it is a direct consequence of modern aggressive asset management. A number of other ratios may be calculated from BS figures using bonds, preferred stock, and common stock, but these calculations are of primary concern to financiers and are of only peripheral interest to operating managers.

5.5 THE INCOME STATEMENT

The income statement is a record of the business done during the past year and its major costs. As with the balance sheet, certain common entries appear on almost all income statements and again we follow the Merrill Lynch listing [1].

Net Sales

The largest source of revenue for the corporation is listed first. Next follow other, smaller revenue sources, such as rents collected on land or buildings and dividends from stock owned in other organizations (see Table 5.6.)

TABLE 5.6   THE AJAX INCOME STATEMENT

AJAX METAL PRODUCTS CORPORATION
STATEMENT OF CONSOLIDATED INCOME AND RETAINED EARNINGS
Year ending December 31 (in thousands of dollars)

Consolidated income statement	1980	1979
Net sales	33,000	30,600
Cost of sales and operating expenses		
Cost of goods sold	24,600	23,052
Depreciation	900	825
Selling and administration expenses	4,200	3,975
Operating profit	3,300	2,748
Other income		
Dividends and interest	150	81
Total income	3,450	2,829
Less interest on bonds	405	405
Income before provision for federal income tax	3,045	2,424
Provision for federal income tax	1,440	1,095
Net profit for year	1,605	1,329
Common shares outstanding	300,000	300,000
Net earnings per share	5.04	4.14

Accumulated retained earnings statement	1980	1979
Balance January 1	3,945	3,066
Net profit for year	1,605	1,329
Total	5,550	4,395
Less dividends paid on preferred stock	90	90
common stock	360	360
Balance December 31	5,100	3,945

Cost of Sales

The largest cost is listed first. This item, cost of goods sold (COGS), shows the cost of converting raw materials or base stock into a finished product ready for sale. Some chemical companies call this the conversion cost. Elements of COGS include raw material delivery costs, labor costs, and factory overhead costs, such as supervision, rent, heat, electricity, supplies, factory maintenance, repairs, and material storage costs. In other words, all costs incurred under the factory roof. The GAAP require that depreciation be included in manufacturing overhead; thus depreciation is included in COGS. Nevertheless, many annual reports show depreciation as a separate item from COGS. Be aware of the specific use of terms such as cost of sales (COS), COGS, gross margin (G. M.), and operating profit.

Depreciation

While there exist other interpretations of depreciation, the most important for tax-liable organizations is that depreciation is a concept established by the income tax laws, to permit a reduction in the taxable income of the organization. Congress has reasoned that firms must replace worn-out plants and machines, and this reduction in taxable income is designed to encourage such replacement.

Selling and Administration Expenses

These costs include sales salaries and commissions, advertising and marketing expenses, distribution costs, administrative salaries, and other administrative expenses.

Operating Profit

Subtracting the cost of goods sold and all other operating expenses, such as depreciation and selling and administrative expenses, from net sales yields the operating profit. This item is also called the operating margin, or gross margin. Gross margin is a rough indicator of the profitability of an enterprise.

$$\text{Gross margin} = \text{net sales} - \text{cost of sales} \qquad (5.3)$$

A reduction in gross margin can occur as a result of price erosion in the marketplace or increased manufacturing costs, to give two examples. Different steps are required to remedy each situation.

Total Income

Adding operating profit to other sources of income yields the total before-tax income.

Net Profit

Fixed charges such as interest paid to bondholders and tax payments must be subtracted from total income to arrive at net profit. Note that payments to bondholders are an obligation and different from dividends to stockholders, which are paid at the option of the board of directors. Net profit is the famous bottom line that holds such fascination for business executives.

5.6 COMMENTS ON THE BOTTOM LINE

Some managers believe that stockholders understand only the bottom line and are unconscious of changes made within the BS and IS to modify this number. Furthermore, some executive compensation packages are tied to the bottom line, and thus strong pressures are generated to manipulate the annual report to place the company in its most favorable light, even when such reports distort reality.

An example that illustrates this point is the method selected to calculate inventory. The last-in, first-out (LIFO) method of placing a value on inventory is universally regarded as providing a more accurate evaluation of the enterprise's current economic condition than the older first-in, first-out (FIFO) procedure. This method is especially accurate in the inflationary economy of the 1970s and 1980s, because LIFO represents the current market cost to replace inventory.

Because LIFO evaluation produces a larger dollar value for inventory owing to inflation, it increases the evaluation of the cost of goods sold built with this inventory, thus reducing the operating margin and the tax burden. Yet LIFO also reduces apparent net earnings, indicating a lower, but more correct, efficiency rating, which may reduce the indicated executive bonus. Thus even though it results in a lower tax burden, many companies do not use the LIFO method.

This is a simplified discussion and it is true, as Granof and Short point out [2], that FIFO does sometimes result in lower taxes. It is also true, however, that GAAP permits use of the accounting procedure that reduces taxes and simultaneously, another procedure for inventory reporting that makes management look good. Specifically, a firm is permitted to show on its income statement for depreciation a figure that has been calculated by the straight-line method, even though an accelerated depreciation method is used to calculate the tax owed. The artificially low reported depreciation number results in an artificially high figure for operating income and an artificially high net profit or bottom line. This strange procedure is also permitted by the Securities and Exchange Commission (SEC).

5.7 EFFICIENCY RATIOS

Several useful ratios are computed from BS data alone, but many other ratios require data from the IS as well. This section considers several of the most important of these ratios.

Inventory Turnover

The proper level of inventory depends on the kind of business engaged in, the season of the year, and the level of the general economy. For example, an automobile dealer wants a high inventory in the spring and a lower inventory as fall approaches.

Inventory in manufacturing enterprises has been brought under increasingly sharp scrutiny in recent times. American manufacturers have much to learn from Japanese organizations in this regard. Japanese managers argue that large inventories of raw material and partially completed products, called work in progress (WIP), indicate sloppy production management. Toyota managers feel that *any* manufacturing inventory indicates management inefficiency! We will have more to say in Chapter 8 about Toyota's famous "just in time" inventory control and the contrasting "just in case" alternative practiced by American production managers, but for now let's calculate the inventory turnover for Ajax and the other American manufacturing firms chosen for comparison.

An approximation of the number of times the inventory has turned over in the preceding year is the ratio of annual sales to year-end inventory.

$$\text{Inventory turnover} = \frac{\text{sales}}{\text{year-end inventory value}} \qquad (5.4)$$

Strictly speaking, one should use COGS in the numerator and the average inventory figure over the year as the denominator, but average figures are not readily available, and thus year-end inventory is universally used. Ajax sales for 1980 are given on the IS as $33,000,000 (see Table 5.6) and the inventory at the BS date is $8,100,000 (see Table 5.3).

$$\text{Ajax 1980 inventory turnover} = \frac{\$33,000}{\$8,100} = 4.07$$

Table 5.7 gives comparative figures for several manufacturing organizations.

TABLE 5.7 ANNUAL INVENTORY TURNOVER FOR FOUR MANUFACTURERS

1980	Inventory Turnover	Inventory Percent LIFO
General Electric	7.47	84
Reynolds Metals	5.40	66
Tandem Computers	5.20	0
Ajax Metal Products	4.07	100

A significant portion of GE's business is the construction and installation of huge steam boilers, electric generators, and nuclear reactors that take two to three years to complete. During the interim these units represent work in progress (WIP) inventory, to which value is constantly added. This situation materially reduces inventory turnover at

GE. Furthermore, LIFO increases the value of the inventory and further reduces apparent turnover. Therefore, the inventory turnover given in Table 5.7 must be the result of intense and relatively successful efforts at inventory control at GE.

In contrast, Tandem is in a fast-moving business and uses FIFO, so a high turnover might be expected. However, Tandem doubled its sales volume from 1979 to 1980, a possible factor in its apparent slow inventory turnover.

If one were to compare GE to the figures given in Westinghouse's annual report, one would see that GE has a fractionally faster turnover. Likewise, Tandem's major competitor, Prime Computer, shows a fractionally slower inventory turnover than Tandem's. Such intra-industry comparisons are generally more meaningful than cross-industry comparisons. Because Ajax is in a relatively simple business, is small, and is not growing rapidly, it should be able to control its inventory better than it is now doing.

Inventory control is discussed in detail in Chapter 8, but to indicate the serious problem the United States faces in international competition, an inventory turnover figure of 10 or more is unknown in American manufacturing concerns. The inventory turnover in a typical American manufacturing concern currently is approximately 6. But Japanese manufacturers regularly achieve annual turnovers of 40 or more.

Next, we come to three very important performance ratios. These are overall measures of the economic efficiency of the firm. Perhaps because they are so important, it has been difficult to avoid tinkering with them to "improve" the precision of what they measure. Unfortunately, this has resulted in confusion and complications in the definitions. In fact, an entire book has been written on how to calculate one of the ratios (ROI) [3].

Return on Investment

The ratio of net return after taxes to the investment made in the business is called ROI. This seems perfectly straightforward until the term *investment* is defined. Pierre du Pont, soon after he took over the Du Pont family gunpowder business and began creating the modern Dupont Corporation, searched for a measure or index of performance for the various sectors of the business. He soon settled on what he called ROI, defined as follows.

$$\text{Du Pont ROI} = \text{margin} \times \text{turnover} = \frac{\text{net income}}{\text{sales}} \times \frac{\text{sales}}{\text{total assets}} \qquad (5.5)$$

Pierre du Pont argued, or rather, stated, because he was not in the habit of arguing with anyone, that this ratio measures the effectiveness with which the firm utilizes its invested assets. It has the further advantage of utilizing data readily available from the annual report. As Du Pont management became known for its effectiveness, and especially after Du Pont bought into General Motors after World War I, the Du Pont ROI came into wide usage and this became the common meaning of the term. In the absence of other information, assume that Du Pont ROI is meant when a business executive uses the term ROI.

Return on Assets

Simple rationalization of Eq. (5.5) shows that sales cancel, therefore the Du Pont ROI is really return on total assets (ROA).

$$\text{Dupont ROI} = \frac{\text{net income}}{\text{total assets}} = \text{ROA} \qquad (5.6)$$

Return on Equity

Return on owners' equity (ROE) measures the efficiency of the current value of the owners' investment in the business. ROE is slightly difficult to calculate, however, because the equation uses the OE from the previous year and subtracts preferred stock factors if applicable.

$$\text{ROE} = \frac{\text{net income} - \text{preferred stock dividend}}{\text{OE, previous year} - \text{preferred stock value}} \qquad (5.7)$$

Tandem's ROI and ROE are both excellent, probably owing in large measure to its rapid sales growth. This is especially true of the ROE because the previous year's OE is used in the denominator, and Tandem's sales doubled from 1979 to 1980. Again Ajax does not fare well in these important indicators. Ajax may be ripe for a takeover raid (see Table 5.8).

TABLE 5.8   RETURN ON INVESTMENT AND OWNERS'
EQUITY FOR SELECTED U.S. MANUFACTURERS

1980	Dupont ROI (ROA) (%)	ROE (%)
General Electric	8.2	20.6
Reynolds Metals	5.8	15.6
Tandem Computers	11.2	33.9
Ajax Metal Products	5.5	14.4

5.8 SOURCES AND USES OF FUNDS

The final statement included in corporate annual reports details the sources and uses of corporate income. This document is designed to show the cash flow of the corporation, which it does moderately well. None of the entries in the sources and uses table are new, thus this explanation can be brief. The Ajax statement in Table 5.9 lists two sources of funds. Other possible sources include income from loans to other companies, sale of assets, and stock dividends.

Depreciation is listed as a source, although it is not new money from the outside. But Depreciation was included in the IS as a cost of doing business. Where did the money go? The enterprise paid it to itself, as the sources table shows, and these funds are available to buy replacement equipment or to invest until such replacement is necessary.

The sum of net income and depreciation is one definition of cash flow. The entries under uses of capital in Table 5.9 should be clear in concept.

The analysis of changes statement is not always prepared, but it is convenient. Table 5.9 indicates that Ajax is moving in the right direction. Note from the IS that both sales and Ajax current assets, except for inventory, increased this year. That inventory decreased while sales increased indicates that Ajax is providing tighter management. Note also that current liabilities are down, including AP. Perhaps Ajax has been too zealous here. It's a good idea not to pay bills until absolutely necessary. In the meantime, leave the float or OPM ("other people's money") in the bank to gather interest.

TABLE 5.9 THE AJAX SOURCES AND USES STATEMENT

AJAX METAL PRODUCTS CORPORATION Statement of sources and uses of funds		1980
Funds were provided by		
Net income	$ 1,605	
Depreciation	900	
Total		$ 2,505
Funds were used for		
Dividends on preferred stock	$ 90	
Dividends on common stock	360	
Plant and equipment	915	
Sundry assets	30	
Total		$ 1,395
Increase in working capital		$ 1,110

Analysis of changes in working capital		1980
Changes in current assets		
Cash	$ 450	
Marketable securities	1,170	
Accounts receivable	300	
Inventories	(900)	
Total		$ 1,020
Changes in current liabilities		
Accounts payable	$ 180	
Notes payable	(450)	
Accrued expenses payable	90	
Federal income tax payable	90	
Total		$ (90)

5.9 SUMMARY

The foregoing analysis has been somewhat caustic regarding management's manipulation of balance sheets and income statements to place a nonobjective view of the firm before its owners. Yet a number of these practices are condoned or encouraged by GAAP and the Financial Accounting Standards Board (FASB). For example, it has been argued that LIFO gives a more realistic evaluation of assets tied up in inventory than FIFO does. Yet GAAP includes a requirement to be conservative in asset accounting. GAAP does not permit reevaluating upward to a higher market value or replacement value PP&E, so why do so for inventory? Assets are recorded on the books at cost, which in an inflationary economy is lower than market or replacement. Why make an exception for inventory?

On the other hand, corporate raiders have been known to buy a company with undervalued assets and then dismember the business by selling off its parts.

These are difficult and controversial questions for which no simple universal answers can be given. Present financial statements are not perfect, but they are much more complete than those permitted in the not-so-distant past. Furthermore, U.S. requirements are more stringent than the disclosure requirements of most other nations. Further progress toward full disclosure is to be encouraged.

EXERCISES

1. Ajax showed an increase in OE of $1,155,000 in 1980. Precisely from where did this come? Obviously, it is an increase in accumulated retained earnings, but what is its exact origin?

2. Table 5.4 gives the 1980 current ratio for several manufacturing firms. Calculate the current ratio for 1985 and for the most recent available annual reports. Comment.

3. Table 5.5 gives the quick ratio for several firms. Update the table entries for 1985 and the most recent year available. Comment.

4. Table 5.7 gives the annual inventory turnover ratio for four manufacturing enterprises for the fiscal year 1980. From the annual reports, calculate these ratios for General Electric, Reynolds Metals, and Tandem Computers for 1985 and for the most recent year available. Comment on the comparisons.

5. Weiss argues that critical financial information on a firm can be gathered by careful study of its annual report [4]. What information is available from the auditor's statement? The analysis of financial condition? Footnotes? Income statement? Balance sheet?

6. Weiss recommends calculating the book value per share. What does this additional ratio indicate? How does this differ from the current ratio?

REFERENCES

1. Merrill Lynch, Pierce, Fenner & Smith, Inc., "How to Read a Financial Report", 4th ed, May 1979.

2. M. H. Granof and D. G. Short, "For Some Companies, FIFO Accounting Makes Sense," *Wall Street Journal,* August 30, 1982, p. 12. Note that the U.S. Treasury carries the nation's gold assets on its books at the FIFO figure of $35 per ounce.

3. R. A. Peters, *ROI,* rev. ed. (New York: Amacom, 1979).

4. G. Weiss, "Reading between the Lines of an Annual Report," *Business Week,* March 23, 1987, pp. 164–165.

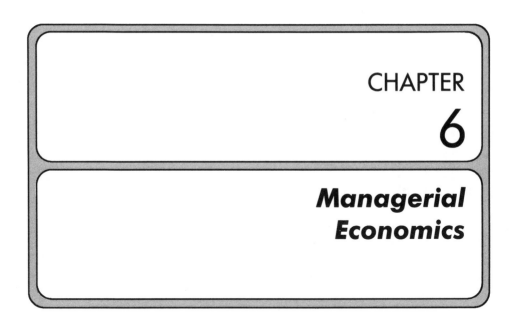

CHAPTER

6

Managerial Economics

6.1 INTRODUCTION

This chapter considers the short-run costs associated with manufacturing a product. The cost of money will be ignored, because the costs involve what is called present economics or immediate economics. Chapter 7 extends the discussion to include a number of strategic financial topics such as the time value of money, capital budgeting, taxation, and long-term performance indices. This chapter approach includes a number of topics usually included in texts on managerial finance and engineering economics. Additional readings may be found in both of these areas.

Computing relevant costs in a manufacturing operation is not simple. Contrary to what one might think, no single correct answer exists to the question of how much a product costs to produce.

6.2 RELEVANT COSTS

Various cost categories can be defined, including direct and indirect, fixed and variable, and short-run and long-run. Remember that these categories and those proposed in other sections of Chapter 6 are established for the convenience of the manager and are defined with various purposes in mind. Chapter 6 focuses on manufacturing costs and the control and reduction of the cost of manufacturing a product. It would be reasonable, however, for a retail sales outlet or a service organization to organize its costs in different categories. With the understanding that the categories are somewhat arbitrary, costs will be divided into the following three major divisions.

Manufacturing Costs

Manufacturing costs are costs incurred under the factory roof and are directly assignable to product manufacture. Manufacturing costs are usually divided into three subdivisions.

1. Direct materials and OEM parts
2. Direct labor
3. Manufacturing overhead

Consider the purpose of each subcategory. After all, it costs money to set up files and to keep records. Why bother if the organization doesn't know what to do with the data after they have been assembled. Materials cost is reflected in the inventory shown on the balance sheet and is often a major cost of doing business. Direct labor represents the value added to the product made and is the source of the unique claim to profit as a manufacturing organization. Manufacturing overhead represents all of those costs incurred in operating the factory, but which are difficult or impossible to assign to a specific product. The costs incurred under the factory roof are all represented on the balance sheet. Raw materials and OEM parts are the incoming inventory, and value added through direct labor and manufacturing overhead are also reflected in the value of the inventory of finished goods on hand.

Selling Expenses

Selling expenses are costs associated with distributing the product. The following costs are included in selling expenses:

- Advertising
- Sales commissions
- Marketing personnel salaries
- Sales promotion

Selling costs are not added to the value of the inventory but are expensed (shown as expenses) in the period, usually the quarter, in which they are incurred.

General Administration

Administrative or front-office expenses include the following:

- Central management salaries
- Office expenses
- All other expenses not tied directly to product

General administrative costs are expensed each period. Remember that these cost categories are set up for the convenience of the management of a manufacturing enterprise and different categories might be more useful in other kinds of organizations. The guiding principle is that all costs should be assignable. That is, each cost should be connected to a specific product, whenever possible and for certain, individual managers must be assigned responsibility for each cost. Obviously, the manager assigned responsibility for a particular cost must also be in a position to control that cost.

6.3 DIRECT AND INDIRECT LABOR

Labor is an important portion of the cost of goods manufactured in the United States. The United States has always been a high-labor-cost nation, and as a result the country has gained many social benefits. But as global marketing becomes the norm, manufacturing managers must reduce labor costs to keep U.S. products competitive in world markets. Conventionally, labor costs are divided into direct and indirect.

Direct Labor

Direct labor costs are those labor costs assignable to a particular product. The traditional method of controlling direct labor costs is the following. First, the manager estimates the hours of labor needed to build an assumed standard number of units at the standard production rate. This number becomes the pro forma labor estimate. After production begins, the actual hours of labor are measured and compared with the pro forma to obtain the labor variance. Labor and other variances are the control variables tracked most closely by managers.

Traditionally, actual direct labor is measured by a device called a job ticket. This form accompanies each production batch through the manufacturing process from station to station. At each workstation the worker enters on the job ticket the start and finish time for the batch and the name of the operation. The more modern approach requires the worker to insert a personal identification badge into a computer terminal reader and to enter through the keyboard the job number, start and finish times, and operation identification number. In this way a computer tracks jobs through the shop and computes daily variances.

Indirect Labor

Labor not directly assignable to a particular product is called indirect labor. Indirect labor costs must be allocated to each product fairly, a task easier said than done. Various ways exist to allocate indirect labor costs among several products. One standard approach is to allocate indirect costs in the same proportion as the direct labor costs incurred. This approach is adequate so long as one product does not absorb a disproportionate share of indirect costs. Suppose, however, that one product uses gold and special guard labor costs are therefore necessary. Although the guards protect the entire plant, they are really only needed for one product, so this product should bear a disproportionate share of the extra guard cost.

6.4 MANUFACTURING OVERHEAD

Manufacturing overhead is the sum of indirect labor costs, indirect materials costs, plant depreciation, and all other unassigned plant costs. We have just defined indirect labor and, by analogy, indirect material costs are those costs of materials utilized under the factory roof but not assignable to a particular product. Items chargeable to manufacturing overhead include product supervision, supervisor salaries, and other indirect labor such as materials handling, security personnel, janitors, quality control, inspection, production engineering, supplies, grease, oil, rags, sweeping compound, factory heat, light, power, and depreciation of PP&E.

In contrast, salaries and office expenses for corporate officers, financial, legal, accounting office expenses, and corporate R&D are not manufacturing overhead but should be charged instead to administrative overhead. This separation may seem simple, but gray areas exist—product development costs, accounting and record keeping done in plant at corporate direction, or computer center costs, for example.

The factory manager says, "Get that charge off my books and over to G. & A.," while the general manager says, "Get those costs allocated out to product lines." Whatever system is finally set up should be adhered to for the sake of continuity and consistency, even though all parties understand that the system is arbitrary in the final analysis.

Note that manufacturing overhead gets absorbed into the value added to the product in the manufacturing process and is represented as part of the value of the inventory on the balance sheet. This procedure is not true of administrative overhead, which is charged to expenses.

6.5 ALLOCATING MANUFACTURING OVERHEAD AT AJAX, SAN ANTONIO

The San Antonio Division of Ajax Metal Products Corporation manufactures fireproof four-drawer files with combination locks. Toward the end of each fiscal year the division cost accountant projects a pro forma budget for the coming year. Based on the current year's operations and at the expected level of production for the coming year, the accountant estimates that $486,480 will be required for manufacturing overhead (see Sec. 6.6 for details). Ajax allocates OH on the basis of direct labor costs, and the pro forma direct labor for the coming year is estimated at $355,000. Thus each dollar of direct labor carries with it approximately $1.37 of manufacturing OH. This number is also called absorbed overhead because it is absorbed into the value of the plant inventory.

In the absence of information to the contrary, the cost accountant assumes that sales and production occur at a constant rate throughout the year. But of course this constant production is quite unlikely in practice. This leads to the important and practical management control concept previously mentioned, called variance. The Ajax fiscal year starts on July 1, and Table 6.1 shows the pro-forma, i.e. preliminary estimates of direct labor and manufacturing overhead, along with the actual situation in the first two months.

TABLE 6.1 VARIANCE BETWEEN ESTIMATED OR ABSORBED OH AND ACTUAL EXPENDITURES AT AJAX, SAN ANTONIO, FOR THE FIRST FEW MONTHS OF THE FISCAL YEAR

	Direct Labor ($)	Absorbed OH ($ estimate)	Actual OH ($)	OH Variance Over (Under) absorption ($)
Estimate for year	355,000	$486,480	—	—
Estimate for month	29,583	40,540	—	—
Actual for July	30,000	41,100	40,000	1,100
Actual for August	30,000	41,100	42,500	(1,400)

If the monthly overhead variances balance out over the year, this method of allocation may be satisfactory, but if a diverging trend develops, we must conclude that actual overhead costs have not been properly allocated by this approach [1]. Generally speaking, this method is unacceptably rough for cost estimation.

6.6 FIXED AND VARIABLE COSTS AT AJAX, SAN ANTONIO

Section 6.5 made some simplified assumptions about manufacturing costs and overhead at Ajax, San Antonio (S.A.) but this section gets a little closer to the actual situation. Ajax S.A., makes three models of fireproof four-drawer filing cabinets. Model A is the stripped version with thinner fireproofing and a cheaper combination lock. Although the gross margin is narrow on this model, Ajax must offer it to meet overseas competition, and it is the company's biggest seller. Model B is the original standard model with thicker fireproofing and a better lock. Model C is the top-of-the-line model. It has the same internal construction as model B and the same combination lock. But model C comes in designer colors and has deluxe front-panel hardware. It does not, however, sport a tiny alligator. Table 6.2 shows the cost breakdown on a full-cost basis as produced by Ajax's cost accountant.

The method of calculating the various items that make up the manufacturing overhead in Table 6.2 is as follows.

Indirect Labor.   Ajax's experience has shown that actual indirect labor costs average about 50 percent of direct labor charges, and the company uses this ratio to produce the numbers in the table. This ratio is typical of industry norms.

Supplies. This item covers all expendable material consumed in the plant. Experience at Ajax, S.A., indicates that it approximates 10 percent of direct labor.

Depreciation—Building. The plant cost $3,600,000 to construct, and Ajax takes a straight-line depreciation over 20 years, or $180,000 per year. To allocate this depreciation on the basis of direct labor, the total pro forma annual direct labor is computed as $355,000, and thus depreciation is 0.507 of direct labor. This ratio was applied to obtain the figures in Table 6.2.

TABLE 6.2 A UNIT COST ANALYSIS FOR AJAX, SAN ANTONIO, CALCULATED ON A FULL-COST BASIS

		AJAX, SAN ANTONIO COST REPORT (Full-cost basis)					
		Model A		Model B		Model C	
Production volume (pro forma)		5000		3000		1000	
Cost elements							
Raw materials and purchased parts		$75		$85		$90	
Direct labor							
fabrication	$10		$15		$20		
assembly	12		20		25		
testing, packing	10		10		15		
		32		45		60	
Manufacturing overhead							
Indirect labor	16.00		22.50		30.00		
Supplies	3.20		4.50		6.00		
Depreciation— building	16.22		22.82		30.40		
Depreciation— equipment	3.50		5.25		7.00		
Utilities and other	4.80		6.75		9.00		
		43.72		61.82		82.40	
Total Unit Cost		$150.72		$191.82		$232.40	

Depreciation—Equipment. The capital equipment is utilized primarily in the fabrication section of the plant, and thus the total annual capital equipment cost of $40,000 will be prorated on fabrication direct labor only. The total annual pro forma direct fabrication labor is calculated as $115,000. Thus equipment depreciation is 35 percent of fabrication labor.

Utilities and Other. These items have an estimated annual cost of $53,250, about 15 percent of pro forma direct labor, and are so allocated in the table.

6.7 CONTRIBUTION

The concept of contribution is of great value as a convenient means of calculating the effect on profit of changes in selling price, production volume, and variable cost items. The drudgery of estimating the effect of such changes has been dramatically reduced by the development of the electronic spreadsheet, and there is no doubt that is the preferred method. Nevertheless, contribution still retains its value as an insight tool for the manager.

Contribution is defined as the difference between the selling price of the product and the variable costs of production. It represents the amount available to cover fixed costs and profit. The terms *marginal contribution* and *contribution to overhead and profit* are synonymous. This concept is valuable because it can be used to focus on the impact of production decisions on the bottom line.

Suppose that the marketing manager at Ajax, S.A., has just recommended a reduction in the selling price of various models with a view to increasing sales. As a first approximation, assume that the suggested production rate changes can be carried out without expensive investments in retooling or in PP&E. How should Ajax go about estimating the effect on profits of these proposed changes? Rotch, Allen, and Smith [1] give several incorrect ways, intentionally, and one or two correct but clumsy approaches, and also the simple method given here. Table 6.3 shows current factory selling price and volume for the three models, along with the proposed new prices and expected new volumes.

TABLE 6.3 CURRENT FACTORY SELLING PRICE AND ANNUAL SALES VOLUME ALONG WITH PROPOSED NEW PRICE AND ESTIMATED NEW ANNUAL VOLUME AT AJAX, SAN ANTONIO

Model	Current Price ($)	Current Volume	Proposed New Price ($)	Estimated New Volume
A	195	5,000	150	12,000
B	240	3,000	200	8,000
C	300	1,000	250	4,000

We could go back now to Table 6.2 and recalculate all of the allocated costs using the new volumes and this is perfectly satisfactory, given a properly set up template on an electronic spreadsheet. But the calculations are tedious by hand. Instead, calculate the unit contribution of each model as shown in Table 6.4.

Compare Table 6.2 with Table 6.4 and note that variable overhead excludes fixed costs such as depreciation, which go on regardless of the number of units produced and includes only items that depend on production rate. To estimate the effect of each change on profit, compute the total marginal contribution as shown in Table 6.5.

TABLE 6.4 THE UNIT CONTRIBUTION OF THE THREE MODELS AT AJAX, SAN
ANTONIO, FOR OLD AND NEW PRICING POLICIES

	Model					
	A		B		C	
Item	Old	New	Old	New	Old	New
Factory price	195	150	240	200	300	250
Less OEM parts and materials	75	75	85	85	90	90
Less variable labor	32	32	45	45	60	60
Less variable overhead						
Indirect labor	16	16	22.50	22.50	30	30
Supplies	3.20	3.20	4.50	4.50	6	6
	126.20	126.20	157.00	157.00	186	186
Contribution	68.80	23.80	83.00	43.00	114.00	64.00

All figures represent dollar amounts.

Overall, the new marketing policy appears successful and should be adopted,
ceteris paribus. The proposed new marketing policy reduces profits on model A but
increases profits for models B and C.

As we examine the situation still more closely, the reason for this becomes clear.
Note from Table 6.2 that model A is priced at only about $20 above factory costs, a very
thin margin. Why then, should Ajax, S.A. bother with model A at all? Why not drop it
from the line? This isn't an unusual problem nowadays for American manufacturers
given the pressure from foreign competition, which is initially most severe on the least
expensive models. We will consider what to do about this in the next section.

TABLE 6.5 TOTAL MARGINAL CONTRIBUTION AS A RESULT OF PROPOSED NEW
PRICING AND MARKETING POLICY DROPS DIRECTLY TO THE BOTTOM LINE AND
REPRESENTS ADDITIONAL PROFIT

	Original Pricing			Proposed Pricing			
Model	Unit Contribution ($)	Volume	Total Contribution ($)	Unit Contribution ($)	Volume	Total Contribution ($)	Marginal Contribution ($)
A	68.80	5,000	344,000	23.80	12,000	285,600	(58,400)
B	83.00	3,000	249,00	43.00	8,000	344,000	95,000
C	114.00	1,000	114,000	64.00	4,000	256,000	142,000
Totals			707,000			885,600	178,600

6.8 RELEVANT PRICING

As described in Sec. 6.7, Ajax, S.A., is faced with a pricing problem. Ajax was forced to cut prices because of competition but to its delight now estimates that the expected increase in volume will actually increase total profit. An exception appears to exist with the repriced model A. It might be argued that because the new model A is priced only $20 above full factory cost, and possibly below total costs, it would be better to drop model A from the line.

Once again, the contribution point of view is the appropriate way to look at the problem. As repriced, model A is expected to produce a total contribution of $285,600 and should be retained. This contribution will be lost if the line is discontinued. Although a full-cost analysis shows that model A is hardly breaking even, this fact is irrelevant. The fixed costs ignored in the contribution analysis are sunk. Ajax, S.A. must pay them regardless of production rates. A new full-cost analysis would assign a greater unit fixed-cost burden to models B and C if model A were dropped.

Does this mean that Ajax is stuck with the reduced profit from model A? Not at all. At least three alternatives present themselves. First, Ajax could seek a replacement product for model A that will utilize the existing PP&E and thus support the fixed costs. The replacement model must produce a greater contribution than model A of course, if the change is to be worthwhile. Second, Ajax could drop model A after it sells off the portion of its PP&E used to produce that model. This action reduces fixed costs and lowers the break-even point. If all of the models are manufactured on the same assembly line, this alternative is foreclosed.

Finally, and most practical in the short run, Ajax should examine the costs of model A with a view to reducing them. Return to Table 6.2 and begin this analysis. The most obvious target for cost reduction is the OEM parts category, because it is the largest single cost factor. A still closer analysis reveals that the major component of OEM costs is the combination lock. Ajax should be able to negotiate a better price from its lock supplier in consideration of the projected substantial increase in volume under the revised schedule. Ajax could also consider substituting a lower-quality lock for further cost reduction.

6.9 BREAK-EVEN POINT REVISITED

The break-even point, mentioned in Sec. 6.8, deserves greater emphasis. Break-even analysis is another of those managerial concepts, such as contribution, that permits managers to slice through mountains of detail to arrive at powerful insights. As with many such ideas, it is simple in concept. Managers realize that fixed costs must be met regardless of production volume. Furthermore, Ajax knows the unit contribution for each of its products. Think of the contribution of each unit of production going completely to meet fixed costs, to that point in the production run at which fixed costs are fully satisfied. At that point the company breaks even, and all further contribution goes directly and completely to profit.

$$\text{Break-even volume} = \frac{\text{annualized fixed costs}}{\text{unit contribution}} \tag{6.1}$$

This equation calculates the break-even point for one product. But in a multiproduct organization it is easy to see that there is no single break-even point. An infinity of combinations of volumes can be chosen that meet all the fixed costs.

As a simple example of a break-even calculation, consider the trash compactor venture in Chapter 4, and assume it must stand alone as a new venture. Take the original selling price of $750, of which Ajax gets $450. Suppose the total operation's annualized fixed costs can be estimated at $3,500,000. Using the same assumptions employed in Sec. 6.6, unit contribution at this selling price is $174. Thus

$$\text{Break-even} = \frac{\$3,500,000}{\$174} = 20,115 \text{ units}$$

This number appears satisfactory on the surface of things, because if Ajax gets 60 percent of an assumed market of 60,000 units, it would sell 36,000 units. But this market share is unrealistic because Sears sells a competitive compactor for $359. Thus the customers value in use isn't $750 but $359. The value in use calculated using normative assumptions is incorrect. Consider Table 4.1, which shows that the standard gross margin in SIC 363 is 19 percent, not the 32 percent received had the $750 figure held true. And, of course, the uselessness of the $750 figure would have been obvious once Ajax checked its competition.

Using a gross margin of 19 percent to calculate a more realistic selling price and assuming that the operating cost estimates are correct, this recalculation results in a retail price of $318.93, which seems much more realistic in light of Sears's price. Assuming a normal retailer's markup, Ajax gets 60 percent of the retail price, or $191.36. Now the unit contribution is negative because the variable costs of production are $276. Ajax can't make money at any volume with this price unless both fixed and variable costs can be reduced dramatically.

Consider a slightly more complex example. Compute the break-even ratio for Ajax's three models of four-drawer files, assuming that the production volumes of the various models maintain the same proportions and prices under the revised schedule. The fixed cost components were used to produce the cost report in Table 6.2. The total of these components is given in Table 6.6.

Note from Table 6.5 that total contribution is $885,600. The pro forma production volume that produces break-even is therefore

$$\text{Ajax break-even ratio} = \frac{\text{Total fixed costs}}{\text{Pro forma contribution}} = \frac{\$273,250}{\$885,600} = 0.31$$

Provided that the various product lines absorb their expected costs, each product line breaks even at 31 percent of pro forma volume. Any other mix of production volumes can be assumed and the break-even point calculated. For example, suppose that model A were dropped and models B and C were produced on the pro forma schedule. The new break-even ratio is 0.48 of the pro forma production volume estimate.

TABLE 6.6 TOTAL ANNUAL
FACTORY FIXED COSTS AT AJAX,
SAN ANTONIO

Depreciation	
Building	$180,000 p.a.
Equipment	40,000
Utilities and other	53,250
Total	$273,250 p.a.

6.10 FULL COSTING VERSUS VARIABLE COSTING

The approach taken in Sec. 6.6, and particularly in Table 6.2, to compute the factory costs at Ajax, San Antonio, is the standard full-costing basis. But the concept of contribution has proved so handy that an alternative costing procedure called variable costing has come into use. Variable costing is said to provide additional managerial insight and depends on variable gross margin, which is simply our old friend contribution under another name.

Under variable costing, the fixed components of manufacturing overhead are removed from manufacturing costs and not included in the value of inventory, but are shown each period as an expense. The arguments for variable costing are the same as those already made for the concept of contribution. Arguments against variable costing charge that it can obscure the very real fixed manufacturing costs and encourages artificially low selling prices for products under attack by competition in the marketplace. Variable costing is not for external use, and figures derived from this approach should not appear in publicly released financial statements.

Compare the abbreviated income statements for Ajax, S.A., for the two costing methods under the original pro forma volume and prices (Table 6.7). The purpose of these calculations determines which approach to use. Suppose, as advocates of variable costing might say, that the effect on profit of a change in sales is desired. The ratio of variable gross margin (contribution) to sales revenue should remain constant under variations in sales volume. This ratio is sometimes erroneously called the profit-volume (PV) ratio. From Table 6.7 note that

$$PV = \frac{\$707,000}{\$1,995,000} = 0.35$$

If sales increase or decrease by $5,000, pretax profits increase or decrease by $5000 \times 0.35 = \$1,771.93$. No such simple relation exists between gross margin in the full-costing analysis and sales volume and profit. However, accounting is complicated by variable costing when inventory changes occur, as is the case when production and sales are not perfectly synchronized.

TABLE 6.7   ABBREVIATED INCOME STATEMENT COMPARING FULL-COSTING AND VARIABLE-COSTING METHODS

	Full Costing ($)	Variable Costing ($)
Abbreviated Income Statement for Ajax, San Antonio (Inventory Level Constant)		
Sales	1,995,000	1,995,000
Variable cost of goods sold	—	707,000
Variable gross margin	—	1,288,000
Fixed manufacturing costs	—	273,250
Full cost of goods sold	980,250	—
Gross margin	1,014,750	1,014,750
Administration and selling	804,000	804,000
Net profit before taxes	210,750	210,750

EXERCISES

1. Table 6.1 uses a ratio of 1.37 times direct labor to estimate the absorbed overhead. This method appears much simpler than the detailed approach in Table 6.2. Is the result approximately the same? Is the similarity accidental? Why or why not?

2. Show that the break-even ratio for Ajax, San Antonio, is 0.48 if the model A filing cabinet is dropped, and the new pricing and production schedule is maintained for models B and C.

3. In the variable-costing approach, the term *profit-volume ratio* is sometimes used. The text calculates this ratio but calls it erroneous. Why? See Ref. [1].

REFERENCE

1. W. Rotch, B. R. Allen, and C. R. Smith, *The Executive's Guide to Management Accounting and Control Systems* (Houston, Texas: Dame, 1978).

Making Long-Term Economic Decisions

7.1 INTRODUCTION

Chapter 7 considers various criteria for judging the long-term economic efficiency of a venture as well as some of the long-term, strategic decisions managers face in maximizing the economic efficiency of a project over time. Appendix A reviews basic calculations involved in determining the time value of specific cash flows.

7.2 THE TIME VALUE OF MONEY

Money can be defined as any common medium of exchange. Money is also an instrument of national and international policy, as when the Federal Reserve System increases or decreases the amount of money in circulation by varying the reserve requirements of its member banks. Money may also be considered a commodity, that is, one of the tools needed to do business. As with other tools, money may be rented.

What elements determine the rental charge, more commonly called interest, a bank gets for the use of its money? Perhaps before beginning this discussion, it should be noted that one need not do the following exercise explicitly. As Milton Friedman would certainly say if asked, the market establishes the interest rate, or more technically, the discount rate a bank charges [1].

The following three factors generally influence the discount rate.

The "Risk-Free" Investment

One form of risk-free investment is a loan to the U.S. government. One can make such a loan by buying a U.S. Treasury bond, which is backed or "secured" by the "full faith and credit of the U.S. government." We put "risk free" in quotes above, because it is not risk free in the most general sense. Governments have been known to repudiate their debts. But by convention the term is in universal use, and U.S. Treasury bonds are the standard of judgment.

Inflation

Inflation is like a habit-forming drug. One administers it carefully and only for the best of reasons, but, as time goes on, it takes a larger and larger dose to create the same effect. Then as one realizes the danger and attempts to cut back, withdrawal symptoms occur. One withdrawal symptom is increased unemployment such as the U.S. experienced in 1981 and 1982. A more serious effect is continued acceleration of the inflation rate leading to total economic chaos, as in the Weimar Republic of Germany following World War I, and Brazil in 1989. Congress and the executive branch generally want to approve more programs than can be supported by available tax revenues. Citizens like programs and dislike taxes. Thus it is not surprising that some elected officials strive to produce one without the other.

Inflation in the United States ran at 4 to 6 percent annually for several decades. Then under the pressure of OPEC energy price increases and national unwillingness to pay for these increased costs in real dollars, inflation crept upward through the Carter presidency. In 1979 inflation increased more rapidly, and in early 1980, depending on the precise measure one chooses to use, the inflation rate hovered between 13 and 17 percent. If risk-free investment has a return rate of 4 to 6 percent, the traditional rate, it is not surprising that the prime rate approached a shocking 20 percent in March 1980.

The prime rate is the interest rate large banks charge their best customers. This rate covers all bank service costs such as bookkeeping costs, computer charges, and salaries for personnel, as well as a reasonable profit. Before the early 1970s the prime rate ran at 4 to 6 percent per year. During the Reagan years inflation dropped off, and the prime rate moved as low as 7.5 percent in early 1987. In 1989 renewed inflation fears began to be expressed.

Risk Premium

The risk premium depends on the experience the lender has had with the borrower and the collateral used to guarantee the loan. The borrower's credit rating is estimated by firms organized for that purpose. The value of collateral depends on the ease with which the lender can capture it and convert it to cash should the borrower fail to repay on time. The risk premium varies from zero for firms able to command the prime rate to several hundred percent for people borrowing from loan sharks. An ordinary business investment entails some risk; thus several percentage points are the standard risk premium. The rate could be considerably higher for entrepreneurs, who are likely to have no credit rating or business track record.

Other

A number of other factors influence interest rates. Actions by the U.S. Federal Reserve Bank that restrict or loosen the credit it extends to member banks have a direct effect on interest rates. Changes in U.S. policy toward selling gold reserves affect domestic interest rates, as does foreign evaluation (by the so-called "Gnomes of Zurich") of the U.S. domestic financial position.

7.3 VARIOUS MEASURES OF ECONOMIC WORTH

It may surprise the reader to learn that the economic evaluation of alternative investment opportunities is a subject not well understood by the managers of today's business enterprises. Not only is the state of ignorance concerning objective economic evaluation extensive, but a great deal of misinformation also exists.

In a survey of research and development venture managers, Schwartz and Vertinsky gathered data on the criteria these managers used to judge the worth of proposed research ventures [2]. Some 47 different indices were mentioned, and six criteria emerged as widely used key factors. In the order of popularity with these practical managers, the six criteria are as follows:

1. A priori probability of overall project success
2. Payback period
3. Return on investment (ROI)
4. Market share
5. Relative cost of project
6. Availability of government funding

Payback period and return on investment are measures of economic efficiency. The other four criteria are measures of risk. Consider these two economic criteria here along with three other popular economic criteria, net present worth (NPW), benefit-cost ratio (BC), and internal rate of return (IRR).

Net Present Worth

NPW is one of several present value measures in current use for calculating the economic worth of business investments; however, the Schwartz and Vertinsky survey finds that it is not among the most popular. Yet research has shown that only NPW can be relied on to maximize the investor's return. It is universally agreed that investor's return is the philosophically correct criterion.

An elementary survey of general business principles such as this text is not an appropriate place to enter into a detailed technical analysis of the error sources of using IRR and measures other than NPW for optimum venture selection. But we cannot totally ignore the matter because according to one survey about 75 percent of the world's commerce is conducted under these erroneous measures [3].

NPW is calculated by taking the difference between the present value of the benefits and the present value of the costs. Its value depends on the discount rate assumed, which is thought by some to be a disadvantage. Yet there appears to be no way of avoiding this assumption (counter claims not withstanding). It will be apparent as well, that the NPW of a project is directly dependent on the total amount of money invested, which may obscure the relative worth of a small, high-quality project hidden among larger projects.

Benefit-Cost Ratio

The BC ratio has been suggested as a way to normalize or unitize a venture on a per-dollar basis so that it may be compared with other opportunities on the basis of the quality of the investment, independent of the quantity of money required. The BC ratio is calculated as the ratio of the present value of the benefits to the present value of the costs.

Return on Investment

In Chapter 5, the concept of ROI was briefly discussed. It was pointed out that the Du Pont definition of ROI makes it equivalent to the return on assets (ROA). Yet in an extensive practical treatment of the concept, Peters, after listing over 20 definitions of ROI in current use, ends by equating it to the internal rate of return (IRR) [4]. IRR is also called rate of return (ROR). It is also possible that individuals may use the term ROI when they really mean net present worth (NPW) defined in this section, or even return on (owners') equity (ROE), defined in Chapter 5. This definitional confusion requires careful definition of the term if one is using it.

Internal Rate of Return

The IRR, also called ROR, is another measure of venture worth in common use. IRR is defined as the hypothetical discount rate at which the present value of benefits equals the present value of costs. This hypothetical discount rate is equivalent to an internal rate of return on the investment. It is argued, falsely, that the higher the ROR, the more valuable must be the venture. Another false argument states that use of the IRR eliminates the need to assume a given discount rate. The specific failure of this method is that it does not properly take into account the reinvestment of interim cash flows.

Payback Period

Payback period is a very simple criterion. It is used even more widely than IRR. Payback period is defined as the length of time required to cause the cash inflow to equal the initial investment. No discounting is involved, and the payback period can be calculated by simple division. Although payback period is not a proper measure of economic efficiency, it is an approximation of risk exposure.

7.4 PRESENT-VALUE CALCULATIONS

The concept of present value is used to reduce the different time streams of cash produced by several projects over varying numbers of years to a single common denominator as it were, to facilitate objective comparison of several opportunities.

Example 7.1 The Cost of the *Denver Post*

Business Week for November 10, 1980, carried an article concerning the purchase of the *Denver Post* newspaper by the Los Angeles Times Mirror conglomerate. The opening paragraph of the article follows.

WHY TIMES MIRROR WANTS DENVER'S *POST*

"We have not lost our marbles," says Otis Chandler, vice-chairman of Times Mirror Co., the Los Angeles publishing and timber products giant. He is reacting to criticism heaped on the company since it announced on Oct. 22 that it was acquiring the floundering *Denver Post* for a surprising $95 million.

The article then says that Chandler is convinced that his company can engineer a quick turnaround and reach a pretax margin of 15 percent in three years. The article goes to say that if Chandler's announced goals are reached, Times Mirror will have paid "a bargain-basement price." It continues:

The terms of the complex deal, widely misreported, call for an upfront payment of $25 million. An additional $55 [million], on which the Times Mirror will make semiannual interest payments at a 10% rate, is due in 1990. The remaining $55 million—interest free—is payable in installments between 1991 and 2000.

The article argues that Times Mirror is counting on continued inflation so that repayment can then be made with inflated, cheaper dollars. It is true that the position of a debtor improves during inflationary times. *Business Week* then gives the opinions of unnamed experts, who indicate that the real value of the deal is $75 million, and says that although "Chandler will not confirm this [he] admits the consensus of the Times Mirror's board put the deal 'in the range of $70 million'."

The article continues, discussing various reactions to the purchase. Note the price mentioned in the first paragraph: $95 million. What is the meaning of this number?

The cash flow mentioned in the story shown in Fig. 7.1 seems to imply that *Business Week* simply added the three principal amounts of $25 million, $55 million, and $15 million to get $95 million. But this is nonsense. It implies that the discount rate is zero for one thing and it ignores the required interest payments between 1980 and 1990 for another. Note from the figure that there are 20 payments of $2.75 million each, if *Business Week* is correctly reporting the 10 percent rate as a nominal annual interest rate. Thus on a zero discount basis,

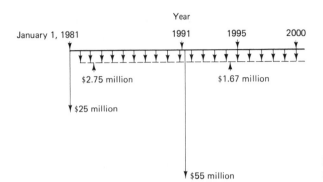

Figure 7.1 Flow of Times Mirror costs as reported by *Business Week.*

the deal is worth not $95 million but $150 million. In reality, the deal is worth neither $95 nor $150 million. Rather, its present worth depends instead on the discount rate, or rates, one assumes will be active over the life of the project.

Chandler is said to be unwilling to confirm an estimate by "experts" of the present value of the costs as $75 million. He does reveal, however, that the consensus of his board puts the project's value "in the range of $70 million."

This incident brings to mind the Indiana legislature, which in 1897 seriously debated simplifying the ratio of the circumference to the diameter of the circle. Some Hoosiers may still feel that pi should be the whole number 4, and Chandler's board may vote by an overwhelming majority that the deal is "worth about $70 million," but in one case nature and in the other the market will decide. But perhaps we are being too stern here. Perhaps what the reporter meant to convey was the board's joint estimate of future discount rates.

Let's take a detailed look at the present worth of the costs the Times Mirror has incurred with this purchase. The functional relationship for the present worth of costs (PWC) may be written as follows (see Appendix B):

$$\text{PWC} = \$25 \text{ million} + \$2.75 \text{ million} \, (P/A, \, i\%/2, \, 20)$$
$$+ \$55 \text{ million} \, (P/F, \, i\%, \, 10)$$
$$+ \$0.75 \text{ million} \, (P/A, \, i\%/2, \, 20) \, (P/F, \, i\%, \, 10)$$

Next, ask two questions. First, what is the most reasonable discount rate to use in estimating the PWC? Second, what is the rate implied by the Times Mirror board when it estimates the worth or cost of this project to be $70 million?

On page 115 of the same issue of *Business Week* the prime rate is given as 14 percent. The next week the prime climbed to 15.5 percent. Experts stated in November 1980 that Reagan and the Federal Reserve had to keep money rates high for the next several years to control inflation. Would 15% be too high a rate to assume for this project? Probably not, because this specific deal might come in several points above prime.

Figure 7.2 plots the present worth of the costs as a function of the discount rate. If the Times Mirror board members think their deal will cost about $70 million, they have implicitly assumed a discount rate of 14.5 percent. *Business Week* experts on the other hand have an implicit rate of 12.5 percent because they placed a worth of $75 million on the transaction. If one feels that the discount rate will vary over time this estimate can be used to influence the calculation. One should also keep in mind that at these high rates, the present worth of a cash item ten or more years away is severely depreciated. At 20 percent, for

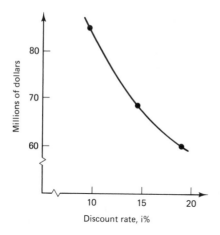

Figure 7.2 Present worth of costs of the Times Mirror deal as a function of the discount rate.

example, Eq. (A.1) in Appendix A tells us that a dollar available in 1999 is worth only 16 cents in 1989, even ignoring inflation.

In 1988 Times Mirror sold the *Denver Post*, taking an after-tax loss of $25.3 million [5].

7.5 THE NET PRESENT WORTH CRITERION

Examine now NPW as the criterion for selecting among possible ventures for funding. The method is illustrated by comparing two energy ventures.

Example 7.2 Two Energy Ventures

Project A. Design and fabrication of photovoltaic cell solar collector units for residential installation. This project is a long-term activity, and forecasts indicate that pretax operating profits, less operating costs, would be $500,000 annually. Capital costs are estimated at $1.5 million initially, $1 million after one year, and $500,000 after two years. Cash flow for the project is shown in Fig. 7.3. The organization requires that all benefit streams considered in venture analyses be limited to ten years. This assumption is common in American business organizations.

Project B. Heat transfer efficiency measurement units for ocean thermal gradient energy recovery devices. Purchase of patent rights cost $500,000. Additional startup costs are $1 million initially and another $1 million after one year. Another $500,000 will be required after two years. The project is to last four years because only a limited number of demonstration units are to be constructed. Operating benefits less operating costs are estimated at $1 million for each of the four years. The estimated cash flow for Project B is shown in Fig. 7.4.

What is the rational way to choose between these two projects? The benefit streams are not the same for the two projects, nor are the time periods. The organization needs to establish a common basis for comparison. The standard basis of comparison is the present value of each of the proposed ventures.

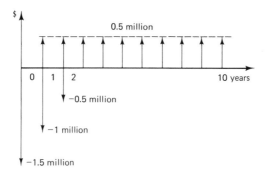

Figure 7.3 Cash flow for Project A.

To calculate the net present worth (NPW) of Project A, use Eq. (A.21) to find the present value of the benefits and subtract the present value of the costs. Note that it is necessary to choose a discount rate to complete the calculation. This adds a judgmental element because one cannot predict future discount rates with perfect certainty. Let $i = 0.08$ for this calculation.

$$\text{NPW}_A = A(P/A, \, i\%, \, n) - F(P/F, \, i\%, \, n)$$
$$\text{NPW}_A = \$500,000 \left\{ \frac{1 - (1.08)^{-10}}{0.08} \right\} - \$1,500,000$$
$$- \$1,000,000 \, (1.08)^{-1} - \$500,000 \, (1.08)^{-2}$$
$$= \$3,355,040.70 - \$2,854,595.34 = \$500,445.36$$

For Project B,

$$\text{NPW}_B = \$1,000,000 \left\{ \frac{1 - (1.08)^{-4}}{0.08} \right\} - \$2,854,595.34$$
$$= \$3,312,126.84 - \$2,854,595.34 = \$457,531.50$$

At the assumed discount rate, Project A is slightly preferable. But do not be content with one calculation at one arbitrary discount rate. Suppose the results are very sensitive to that rate and the rate changes. Because the decision may critically depend on the assumed discount rate, what can be done? One prudent solution is shown in Fig. 7.5, which plots the NPW of the two projects as a function of the discount rate.

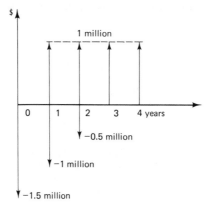

Figure 7.4 Cash flow for Project B.

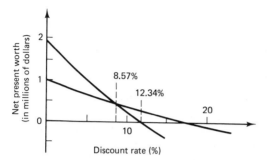

Figure 7.5 Projects A and B as a function of the discount rate.

While Project A has the greater NPW at lower discount rates, Project B is superior at higher rates. The crossover occurs at 8.57 percent; thus the choice of which project to recommend is critically dependent on the assumed discount rate.

Consider another point. If the organization recommended Project A without qualification, a hidden assumption would have been operating. Hidden assumptions are a constant plague, and managers must train themselves to be wary of them. For Project A, the hidden assumption is that the effective discount rate over the next ten (not four) years will be less than 8.57 pecent. In this example, the hidden assumption was not hard to find. Most are not as obvious.

7.6 BENEFIT-COST RATIO

Calculating the NPW of two energy ventures in Sec. 7.5 for a discount rate of 8 percent resulted in a preference for Project A over Project B. Apply the BC criterion to the same two projects and see if the same result holds.

Example 7.3

  Project A

$$BC = \frac{\$3,355,040.70}{\$2,854,595.34} = 1.1753$$

  Project B

$$BC = \frac{\$3,312,126.84}{\$2,854,595.34} = 1.1603$$

Project A appears slightly superior to B. The results aren't surprising because the BC ratio is the same as the NPW criterion on a per-dollar-of-cost basis. The BC ratio is sometimes used in the private sector, but it is more popular for evaluating the relative worth of projects in the public sector.

7.7 INTERNAL RATE OF RETURN

Examine Fig. 7.5 and note that the PW of benefits equals the PW of costs at 12.34 percent for Project A and at approximately 18 percent for Project B. That is, at each of these discount rates, the NPW of the respective venture is zero.

This calculation is the IRR criterion defined earlier, and it seems to indicate that Project B is not just superior to Project A but greatly so. This margin of superiority may be somewhat misleading, however. In Fig. 7.5, careful examination of the shallow slope at which the curve for Project B cuts the axis shows that an incremental change in the original estimate of costs affects the IRR of Project B more than that of Project A.

Project B is more sensitive to cost variations than Project A. This sensitivity is important because benefit and cost estimates are only educated guesses. In the managerial decision process, the question of sensitivity to reasonable variation in any of the critical parameters cannot be ignored.

7.8 THE PAYBACK PERIOD CRITERION

It was mentioned earlier that the IRR criterion is more widely used than the more rigorous NPW criterion. The payback period criterion is more widely used than even IRR, and with less theoretical justification. Payback period is defined as the length of time required to cause the cash inflow to equal the initial investment. No discounting is involved, and the value may be calculated by simple division.

Example 7.4

Compute the payback period for Venture A and then for Venture B (Figs. 7.3 and 7.4).

$$\text{Project A payback period} = \frac{\$3,000,000 \text{ original investment}}{\$500,000 \text{ benefits/year}} = 6 \text{ years}$$

$$\text{Project B payback period} = \frac{\$3,000,000 \text{ original investment}}{\$1,000,000 \text{ benefits/year}} = 3 \text{ years}$$

By the payback period criterion, Project B is markedly superior. The payback period criterion avoids the difficult calculations needed for NPW calculations and the elaborate trial-and-error processes to discover the IRR. The reader may feel we have finally struck paydirt. Have we saved the best till last? Does simple division do the job? Simple or not, it should strike us as a matter of concern that the payback period criterion indicates that Project B is optimum whereas previously Project A had been chosen by the NPW criterion and the BC ratio.

The actions of practical business people say that the payback period is a meaningful criterion, while theorists say it is not. Business people may find the simplicity of the payback period criterion appealing, but the criterion would fall into disuse if it produced misleading results. Because it is widely used, perhaps it should not be dismissed out of hand. Bussey argues that the payback period is used for three main reasons.

First, payback period is not merely a means of estimating economic return. It is also a means of measuring risk exposure. In all our analyses thus far, we have assumed that the future costs and benefits are definitely and unequivocally known in advance. This is obviously not true. We do *not* know the future without doubt. This doubt or uncertainty adds risk to our assessment of the future economic return. Clearly the less time our investment is at risk the better. Thus a short payback period is good because it minimizes risk, not merely because it indicates a good return.

Second, it can be shown that for the simple case of an investment at $t = 0$ and a constant return each year for n years, the payback period is the reciprocal of a crude measure of the equivalent rate of return on the investment. If this effective interest rate on the periodic declining investment balance is greater than the firm's marginal attractive rate of return, it signals an attractive opportunity.

Third, in the typical situation of limited capital available to the firm, payback period indicates to the manager, roughly when the investment will be recovered and available for other use. [5]

Thus payback period is a rough measure of several rather important attributes of an investment opportunity. But it is an approximation at best. The example to which section 7.9 is devoted is designed to bring to a head the problem of which criterion is the proper one to use in selecting from among several venture offerings.

7.9 A CRITICAL EXAMPLE

This example has been designed to show that the optimization criterion affects the outcome of the selection process. The four popular criteria are applied to each of four cash flows. Each criterion indicates a different project as the most economically efficient choice. Because all four cannot be optimum, this demonstrates that it is essential to select the proper optimization criterion. Suppose the following four investment opportunities have been presented from which to select the best option in which to invest internal funds.

Project W.   A small project that requires a Unix-compatible computer software package is to be made available for a stand-alone microprocessor installation for use over the coming year. The package can be leased for $100 for the year and will yield an estimated $115 in annual benefits. At the end of the year the design office is to be converted to a distributed processing system with a standard software set. No future use is anticipated for the leased software package. (See Figure 7.6.)

Project X.   The engineering design office requests the purchase of a used car for in-town courier service. The auto costs $3000 and is expected to return $1500 net benefits each year for three years. The machine will be junked at that time and will have no salvage value. (See Figure 7.7.)

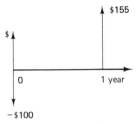

Project W

Figure 7.6   Cash flow for Project W.

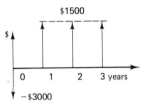

Project X

Figure 7.7   Cash flow for Project X.

Project Y.   The bank has proposed a rather unusual long-term investment opportunity. For a $500 certificate purchased now, it will pay $12,295 at the end of 20 years. Evaluate this unit investment for the firm. (See Figure 7.8.)

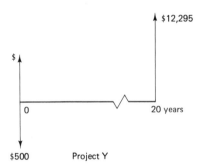

Project Y

Figure 7.8   Cash flow for Project Y.

Project Z.   A new stamping press to carry peak loads in the sheet-metal shop costs $5000 initially and will return $1200 net revenue annually for ten years. It will then be junked with no salvage value expected. (See Figure 7.9.)

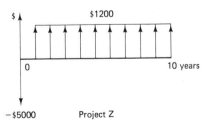

Project Z

Figure 7.9   Cash flow for Project Z.

The company comptroller has decreed that all investments be evaluated over their lifetime at a 15 percent discount rate. Evaluate these four projects for maximum economic efficiency by IRR, BC ratio, NPW, and payback period.

Project W: Microprocessor Software Lease

The rate of return is defined as the discount rate for which the present value of benefits equals the present value of costs. Thus this project seeks i such that

$$\$100 = \$115(P/F, i\%, 1)$$

Generally, a trial-and-error approach to this calculation is required, although this case happens to be particularly simple. Note that an interest rate of $i = 15$ percent satisfies the equation.

$$\text{IRR} = i = 15\%$$

The benefit cost ratio at $i = 15$ percent is found as follows:

$$\text{BC} = \frac{\$115\ (P/F,\ 15\%,\ 1)}{\$100} = \frac{\$100}{\$100} = 1.0$$

The NPW is given by

$$\text{NPW} = \text{NPB} - \text{NPC} = \$100 - \$100 = 0$$

If a smooth curve is drawn to represent instantaneous cash flow, the payback period is slightly less than 0.5 years (exactly 0.465 years). Such an interpolation process doesn't make sense, however, because the organization is operating under an accounting convention that permits cash transfers to be made only at the end of each interest period. For these projects the interest period is at year-end. Thus the payback period is one year.

Project X: Courier Service Used Auto Purchase

The IRR for Project X occurs at an i for which the following relation holds.

$$\$3000 = \$1500\ (P/A,\ i\%,\ 3)$$

Trial and error will produce a solution for this equation at

$$\text{IRR} = i = 23.3\%$$

NPW at $i = 15$ percent may be found as follows:

$$\text{NPW} = -\$3000 + \$1500\ (P/A,\ 15\%,\ 3) = -\$3000 + \$1500(2.283)$$
$$\text{NPW} = \$424.50$$

And

$$\text{BC} = \frac{\$1500(2.283)}{\$3000} = 1.14$$

The payback for Project X occurs in two years.

Project Y: Certificate of Deposit

In the same fashion, calculate the following:

$$IRR = i = 17.3\%$$

$$NPW = -\$500 + \$12,295 \ (P/F, 15\%, 20)$$
$$= -\$500 + \$12,295(0.0611) = -\$500 + \$751 = \$251$$

$$BC = \frac{\$751}{\$500} = 1.5$$

Payback period = 20 years (not 0.8 year)

Project Z: Stamping Press

$$IRR = 20.2\%$$
$$NPW = \$1022$$
$$BC = 1.2$$
$$Payback = 5 \ years$$

In Table 7.1 the results are given with the best choice for each criterion indicated by a square bracket. In this example there are four projects, each of which is optimally efficient according to one criterion and nonoptimum in any other sense. This admittedly artificial example illustrates the confusion among the various criteria. Which is the "correct" criterion and what does "correct" mean? This example does not indicate which is the correct result. However, it is universally agreed by economists that NPW is the fundamentally correct criterion and therefore Project Z is optimum for economic efficiency [6].

TABLE 7.1 FOUR PROJECTS EVALUATED FOR ECONOMIC EFFICIENCY BY FOUR COMMON CRITERIA. (Each criterion indicates a different optimum project, shown bracketed [ ]. Net annual worth is to be preferred to NPW in comparing projects with different lifetimes, but that does not change the result in this example.)

Project	Payback (years)	IRR (%)	BC	NPW ($)	Net Annual Worth ($)
W	[1]	15	1.0	0	0
X	2	[23.3]	1.14	424.50	185.70
Y	20	17.3	[1.5]	251.00	40.10
Z	5	20.2	1.2	[1022.00]	[203.90]

Note that the form of the cash flow in each of these projects is particularly simple, consisting of a single investment at time zero followed by a flow of benefits. This illustrates that the results are not due to "nonnormal" cash flows as is sometimes erroneously claimed. Nevertheless, there are other concerns with this example. It might be argued that the unequal project lifetimes could invalidate the result. This is a valid objection. However, an annual net worth (ANW) that is equivalent to NPW can always be computed and ANW places the comparison on a fair annual basis. This does not change the result here. In general, it is necessary to establish a common length of time over which to make comparisons among ventures.

Any possible concern with unequal lifetimes is eliminated in the example given in Table 7.2. In this example there are two cash flows over the same lifetimes that require the same initial investment and for which NPW and IRR criteria differ as to the proper choice.

TABLE 7.2   TWO CASH FLOWS OVER THE SAME PERIOD OF TIME AND REQUIRING THE SAME INITIAL INVESTMENT (The NPW criterion and IRR differ as to which project maximizes economic efficiency.)

Project	Initial Investment ($)	EOY 1	EOY 2	EOY 3	EOY 4	EOY 5	NPW ($)	IRR (%)
A	− 10,000	3,000	3,000	3,000	3,000	6,000	[1,548]	20.7
B	− 10,000	7,000	7,000	500	500	− 2,000	1,000	[23.9]

Note: Optimum project bracketed [ ].

Another point that must be clear in constructing a truly definitive and rigorous example is the mutual independence of the projects, as opposed to their possible mutual exclusiveness. Still another limitation in these examples is the assumption of a fixed MARR over the lifetimes of the projects. Moreover, it must be known if the amount of available capital can limit the number of projects to be selected. To develop more rigorously these and other points of possible concern would require several chapters. The interested reader is referred to the texts by Au and Au, and Bussey previously mentioned.

7.10 THE CAPITAL-BUDGETING DECISION

So far in this chapter we have completed only one part of the job we started to do, which is to illustrate the "typical" capital budgeting decision. Various economic selection criteria have been discussed and it has been shown that the choice of a particular criterion can influence the result of the selection process. Now we move on to several other important considerations in capital budgeting.

The first consideration is the matter of depreciation, and the second is the influence of taxes on the process. Ideally, a tax system should be so constructed that business decisions are made independent of its effects. Unfortunately, this situation is far from true in the United States. Not only is the playing field not level by happenstance, but also a number of tax and depreciation provisions have been deliberately enacted to influence business decisions.

Let one example from a multitude suffice. A corporation that owns stock in another corporation must pay taxes on only 30 percent of the dividends received on the stock. If the corporation is in the 40 to 50 percent tax bracket, as is common, it pays taxes at a maximum rate of 15 percent on the dividend income. Suppose that you are the corporation's plant manager, and you have submitted a capital budget request. The request must show an ROI superior to the opportunity to invest in another company's preferred stock. Your company pays taxes on the income from investment in your project at a 50 percent rate and only at a 15 percent rate in the alternate opportunity. This thinking may have influenced U.S. Steel (USX) when it made a heavy investment in Marathon Oil rather than investing capital in improving the efficiency of its steel plants. At that time the rate was only 15 percent of the dividends received.

7.11 DEPRECIATION

Although the term *depreciation* can have several meanings, it is used in business almost exclusively in its taxation sense. Standard IRS-approved methods of calculating depreciation are available in all standard texts on business and engineering economics [7]. The following two points are important with regard to depreciaton.

1. Depreciation is purely a bookkeeping transaction used to determine a smaller taxable income than would otherwise exist. It does not represent an actual cash flow.
2. The operating principle is, *ceteris paribus*, to take the largest possible depreciation as quickly as allowed by law to reduce the firm's tax obligation to its lowest possible value in the shortest time. Because of the time value of money, immediate tax savings are more valuable than future savings.

Chapter 5 explained that depreciation appears on the annual income statement of the firm. As pointed out in Section 5.6, however, the depreciation shown on the IS can be without real meaning. Although the most common methods of calculating depreciation are straight line, double declining, and SOYD, there are other approved methods that can be advantageous under special circumstances. Depletion plays an analogous role to depreciation for nonrenewable natural resources such as coal, oil, and other minerals. Amortization is a financial method used to retire debt obligations of the firm.

7.12 THE CORPORATE INCOME TAX

This section considers only the corporate income tax of a firm in ordinary business operation. It ignores carryback and carry-forward provisions, tax losses, write-offs, and other taxes such as value-added and excise taxes.

The first objective is to determine the firm's taxable income. It is found by the following general procedure:

$$\text{Gross income} - \text{operating expenses} = \text{taxable income}$$

Each of the categories on the left side of the equation consists of several important components.

Gross income = gross sales + interest + dividends + royalities + rents

Operating expenses = cash expenses + noncash expenses + interest paid on indebtedness

Cash and noncash expenses may be further broken down.

Cash expenses = labor + materials + indirect costs + losses

Noncash expenses = depreciation + depletion + amortization

Once taxable income has been calculated, the applicable tax rate multiplier is applied to compute ordinary income tax. The corporate income tax schedule is simpler than the personal income tax schedule, consisting of only two or three categories. Roughly speaking, the corporate tax rate is 25 percent for the first $50,000 of taxable income and 50 percent for the remainder.

There is no general rule for estimating the precise effect of income tax on the return on an investment. If the company is profitable and has no extraordinary items to carry back or forward, it is apparent that the income tax substantially reduces the ROI. But the reduction depends on many factors, and it need not be precisely proportional.

Suppose that a company is considering two projects that offer the same before-tax return. If one venture involves major capital depreciation and the other does not, the venture with depreciation is expected to prove superior after tax considerations are included. Thus the company is placed in the position of choosing one venture over the other on the basis of taxes rather than long-term organizational benefit.

Given two ventures, equal before taxes and with the same capital requirements, a company is pushed toward the venture with the shorter depreciation lifetime. Factory buildings depreciate over 45 years, according to the IRS, while trucks and autos do so during a five-year period. Thus the typical industrial firm might hesitate to build a new factory but be lavish with company cars. Yet it seems obvious that a needed new factory building contributes more to increased employment and the financial health of the nation than an equivalent investment in executive limousines.

EXERCISES

1. True or false? In evaluating several alternatives, the use of present worth analysis, annual cash flow analysis, and rate of return analysis yields a consistent ranking of alternatives.

2. A corporation is considering making a gift to a university to endow a chair. How much will the corporation have to give if the chair is designed to pay $5000 annually to the recipient, if the interest rate is 9 percent, if and the chair is to exist for ten years?

3. A special lathe was designed and built for $80,000. It is estimated that the lathe will result in production cost savings of $15,000 per year for 15 years.

a. Assuming a salvage value of zero at the end of 15 years, what is the expected rate of return?

b. What if the lathe becomes inadequate after six years of use and is sold for $20,000? What then is the actual rate of return?

REFERENCES

1. Milton Friedman and Rose Friedman, *Free to Choose* (New York: Avon, 1981). Milton Friedman, well-known American economist and Nobel laureate, is a strong advocate of the principle of the free marketplace.

2. S. L. Schwartz and I. Vertinsky, "Multi-attribute Investment Decision: A Study of R and D Project Selection," *Management Science*, 24, no. 3 (November 1977), pp. 285–301.

3. T. Au and T. P. Au, *Engineering Economics for Capital Investment*, (Boston: Allyn and Bacon, 1983), p. 202.

4. R. A. Peters, *ROI*, rev. ed. (New York: Amacom, 1979).

5. *Business Week*, June 20, 1988, p. 71.

6. J. H. Lorie and L. J. Savage, "Three Problems in Rationing Capital," *Journal of Business*, 28, no. 4 (October 1955), pp. 229–239.

7. L. E. Bussey, *The Economic Analysis of Industrial Projects* (Englewood Cliffs, N.J.: Prentice-Hall, Inc., 1978).

PART

III

Management of Manufacturing Operations

Part III introduces a number of topics commonly identified with the subject of manufacturing engineering. However, the material does not take the conventional manufacturing engineering approach. The conventional bottom-up approach emphasizes the accumulation of detailed techniques and postpones an overall grasp. It is precisely this overall grasp that the manager of the manufacturing enterprise needs. Thus Part III gives heavy emphasis to financial accountability and considers an integrated manufacturing systems perspective using a top-down approach.

CHAPTER
8

Control of the Inventory Investment

8.1 INTRODUCTION

Few topics in the field of industrial production managment in the United States have seen a greater change in the conventional wisdom in the 1980s than the areas of quality control and inventory management. Few manufacturing specialists question the recent heavy emphasis on quality control in the production process, because quality control appeals to the craftsman instinct common among us. Some people, however, have trouble initially with the idea of minimizing inventories. Conventionally trained American manufacturing specialists like high inventories. They like high-raw-material and high-OEM stocks because they provide protection from vagaries in the supply chain. Manufacturers like to be in control of the work situation and don't like to be dependent on truckers or railroads to deliver material needed to operate the plant for the next few days.

The conventionally trained manufacturing manager likes to have small cushions of partially assembled parts between each workstation. These cushions, along with the parts actually in the machines being worked on, constitute the work-in-progress (WIP) inventory. The manufacturing manager sees this parts cushion as desirable because, if an individual machine breaks down, the rest of the line still has enough slack to permit it to continue running until the broken tool is replaced or a replacement machine is moved into position.

Finally, it is conventional wisdom to want a good supply of finished goods on hand to fill any reasonable purchase order the day it is received.

But by current thinking all of these arguments are wrong, or at least often lead to unreasonable expense. Furthermore, advocates of the "just in time" (JIT), inventory control concept argue that the "just in case" style of inventory stockpiling just described not only is expensive but also tends to obscure inefficiencies and incipient breakdowns in the production process while they are still minor and easily correctable. Failure to catch minor problems in a timely fashion can lead to major breakdowns later on, breakdowns that can be extraordinarily expensive and time consuming to correct. Thus the JIT approach should not be viewed as contrary to good production practice but as conducive to excellent practice.

As a consequence of the conventional loose thinking, total inventory constitutes the second largest asset category for all U.S. manufacturing firms, exceeded only by PP&E and followed by receivables. Note from the BS in Chapter 5 that Ajax is typical in this regard.

Furthermore, the annual Economic Report of the President shows that despite all recent talk about tightening manufacturing operations in the United States, the ratio of manufacturing inventories to final sales has remained essentially constant for 30 years (see Table 8.1). Armstrong quotes an authoritative study to the effect that as few as 25 percent of the elaborate computerized materials requirement planning (MRP) systems have achieved their targeted objectives. Armstrong cites another study that recently uncovered the embarrassing fact that 23 errors in the mathematical equations theorists use for modeling inventories have existed in   literature since 1965   The implication of all this seems to be that talk is more common than effective action in inventory control.

TABLE 8.1 RATIO OF
NONFARM INVENTORIES
TO FINAL SALES, IN
CONSTANT DOLLARS

Year	Ratio
1954	2.64
1964	2.61
1974	3.11
1984	2.64

Source: Economic Report of
the President, 1954, 1964,
1974, 1984

The Japanese style of production practice is actually a distillation of the best of past American practice [2]. A specific example can be cited of JIT inventory control. Here is what Richard T. Lindgren, CEO of Cross and Trecker, a well-known American machine tool manufacturer says. Lindgren, a former Ford Motor Company manager, talks about the way inventory was managed at the old Ford River Rouge plant in Detroit:

> Thirty years ago when I worked for Ford Motor at the Rouge, if we didn't get a shipment of sheet metal by noon, we didn't run in the afternoon because we never kept more than a half a day's inventory of steel in the plant at any time. It's what the Japanese call Kanban or "just in time" delivery of supplies. They don't have machinery that's any more advanced than ours. They're just more careful about inventory. [3]

To refer back more than 30 years to the Rouge is not to minimize the work of Japanese managers as imitative but rather to criticize the generation of American production managers who in the recent past, by failing to develop a conceptual, theoretical grasp of their field and obsessively focusing on minutiae, lost the forest for the trees. Successful implementation of JIT inventory control requires a complete rethinking of the manufacturing process. It is not simply another computer control procedure or another set of paper forms labeled "inventory control" superimposed on conventional plant practice. Successful implementation of just-in-time inventory control requires redesign of the entire procurement and manufacturing process. It usually takes more than five years to make this change completely operational.

8.2 IMPACT OF EXCESS INVENTORY ON OPERATING STATEMENTS

Section 5.7 defined inventory turnover as the ratio of the dollar value of annual sales to the dollar value of the year-end inventory. Inventory turn is a simple indicator of the efficiency of the manufacturing operation and, as commented in Chapter 5, it currently shows sloppy practice in the United States. Here is what Lindgren of Cross and Trecker says in the *Forbes* article just quoted in comparing plants he operated in Germany, the United States, and Japan.

> In my last job I ran a company that made asphalt compactors: steamrollers. We had a very well-run company in Germany, and it turned its inventory 3 to 3½ times a year. The company we had in the U.S. wasn't nearly as well run, so we used to get a 2½ turn in our inventory. Stuff was piled up all over the shop. But we were also running a company in Japan, and there we were able to get an 8½ turn in our inventory. That was the difference. [3]

Forbes didn't ask, and Lindgren didn't say, what happened to the U.S. plant manager who couldn't control his inventory, but few people would envy the plant manager's position. Heavy, low technology goods generally show a slower inventory turn than lighter, high technology items, but in the latter the Japanese also excel. In the U.S. electronics industry an inventory turn of 8 or 10 is considered good. We saw in Chapter 5 that in 1980 Tandem Computers and Prime had inventory turns of about 5. Market analysts consider these companies outstanding minicomputer manufacturers and representative of best American high-tech practice, yet they turn over their inventory at a slower rate than a Japanese steamroller manufacturer! In Japan electronic manufacturers regularly show inventory turns of 30, and turns approaching 40 are not unknown.

As Chapter 5 explains, inventory is one of the basic assets of the firm, and its value is declared on the balance sheet (see Fig. 5.1 for example). Why is the number of times the inventory turns over in a year considered to be an indicator of manufacturing efficiency? First, because it is impossible to have a high inventory turnover without outstanding execution of all elements of good manufacturing practice, and, second, a high turnover demonstrates an effective use of corporate assets. Consider this second point in more detail.

The use of inventory turnover as a measure of manufacturing efficiency is not new in the United States. In fact, Alfred P. Sloan, Jr., cites the lack of inventory control at General Motors as one of the causes of its dangerous slide in the depression of 1920. Auto sales were cut in half, and the young corporation might have floundered had Sloan and other corporate officers not forced inventory discipline on the operating divisions. Even though sales were dropping, Sloan forced GM to improve its inventory turnover from 2 turns per year to 4 times annually from 1920 to 1921.

Table 8.2 shows comparative statistics on inventory as a proportion of total assets for several companies. Note that Ajax is doing rather poorly, although the Ajax ratio was considered representative of best American practice a decade ago. Suppose that Ajax were able to reduce its inventory to 15 percent of assets instead of the existing 28 percent by increasing inventory turns. This action would reduce Ajax's total assets to $24.7 million and inventory to $3.7 million. The reduction of assets by $3.3 million would permit an equivalent reduction in liabilities. If Ajax is paying 10 percent on its bank loans, it could save $370,000 interest annually by this reduction. This savings represents a more than 20 percent improvement in earnings and an additional dollar per share for stockholders. Although it is unlikely that all of the inventory reduction would show up on the bottom line, a significant proportion would.

TABLE 8.2 PROPORTION OF CORPORATE
ASSETS IN INVENTORY IN 1980

Company	Proportion of Assets in Inventory (%)
General Electric	17
Prime Computers	15
Tandem Computers	21
Ajax Metal Products	28

Source: Annual reports.

Return on investment is a common measure of managerial efficiency and a ratio to which corporate bonuses are often tied. ROI can be improved by increasing sales or by reducing the investment in assets. Reduced inventory investment is an example of the latter.

8.3 REDUCING RAW MATERIAL AND OEM STOCK

One does not improve the incoming inventory situation simply by reduction by fiat of the established standard backlog quantities. Presumably, the historical levels of OEM parts and raw materials kept on hand were developed as a result of experience. Perhaps a strike at a parts supplier some time in the past caused the firm to increase OEM backlogs. The new Chrysler Corporation has been an enthusiastic leader in reducing inventory stocks. Yet a wildcat strike in October 1983 at a single Chrysler stamping plant threatened to shut down assembly operations throughout the company within three days and must have given the corporation pause in its enthusiasm for JIT inventory control. Sometimes backlogs are needed to cover for unreliable transportation service, and quite often excessive backlogs can be traced to a failure to understand the cost of a just-in-case attitude on the part of the manufacturing management team.

The purchasing department, production engineering, and manufacturing operations managers need to work together in taking a new look at the incoming inventory problem. Expect resistance from all three groups when inventory reductions are proposed. Changes will not occur over night. These internal management discussions will likely result in the conclusion that improved reliability of subcontractor deliveries and quality of incoming parts are necessary if the raw material and OEM backlog is to be reduced.

How can a firm get its suppliers to guarantee on-time delivery of quality-controlled parts and supplies? Obviously, the traditional at-arms-length style of purchasing departments in the United States in dealing with suppliers must change. If a purchasing department concentrates exclusively on lowest first cost in negotiating contracts with suppliers, that is exactly what it will get, and no more. The firm then faces increased incoming inspection costs, a higher rate of parts rejection in assembly, higher rework costs, and generally poor finished-product reliability.

After purchasing people understand what is needed, the company can begin talking with its suppliers. In fact, some large corporations, such as General Motors, find it worthwhile to run short courses to teach their regular suppliers what is expected of them in the future. One typical outcome of this process is reduction in the number of suppliers, and larger purchase orders covering longer periods of time are negotiated. Incoming inspection of OEM parts should be eliminated over time with buildup of a group of certified suppliers. The certified suppliers enjoy a special relationship with the firm, and our quality control people should have regular entree to certified suppliers' factories.

Do not underestimate the difficulty of rooting out old thought habits, however. Even GM, an American industrial leader at inventory reduction, has difficulty getting purchasing managers to trust suppliers. They talk JIT but refuse to negotiate exclusive, long-term contracts. And who can blame them? Whose job is on the line if a supplier fails to meet commitments? Why not have another supplier in reserve, to build up a little cushion? Why not try for a little better price by splitting the contract and pitting one supplier against another? And back they slip to the bad old ways.

In exchange for assured, long-term contracts, the certified supplier must not only take responsibility for quality control of the product and meet the purchaser's standards but also guarantee delivery of parts and supplies on a daily or even hourly basis, as if the supplier were operating one of the purchaser's own stockrooms.

At first suppliers may attempt to provide the new, higher-level of service using brute force and old methods. But they soon discover that this method is unprofitable. Suppliers cannot operate their factories in the same sloppy way and expect to "inspect quality in" at the end of the line. Furthermore, suppliers cannot afford simply to take over the warehousing function. They need to tie their production and delivery schedules closely to the purchasing company's.

Manufacturers need to open their order books to suppliers so that they know daily parts requirements months in advance. Furthermore, manufacturers need to work with suppliers to show them how to improve their quality control. Finally, suppliers need to think about relocating plants close to manufacturers to reduce the pipeline effect and the risk of late parts delivery.

Manufacturers might need to lend suppliers money or buy an equity position in their firms to provide the capital needed to make the necessary changes in production and delivery procedures and qualify as certified suppliers. IBM was among the first American firms to think all this through. This line of reasoning may explain IBM's purchase of an equity position in Intel, one of IBM's major chip suppliers.

This is not the traditional way in which American business deals with suppliers and sub-contractors. Just-in-time inventory control is a whole new way of doing business. It means making partners of suppliers. In Japan suppliers are thought of as part of the family, not quite as close as the company's employees perhaps, but people to whom special, long-term obligations are due. A manufacturer has to be sure that suppliers are profitable if the manufacturer is profitable and likewise expect them to share the burden when profits fall.

8.4 PIPELINE EFFECTS IN INVENTORY CONTROL

Generally speaking, pipeline effect means more than the cost of inventory in transit from one plant to another or from a supplier to the plant, although that cost alone can be large. Say that the ship that carries finished autos or partially assembled subunits from Japan to the United States takes two weeks on the ocean, and another two weeks is consumed in loading and unloading; add over-the-road transport at each end, and you realize that Japanese finished-auto production is subjected to more than a month of interest charges on the cost of finished goods. At current production rates and interest charges, this represents a cost of several hundreds of millions of dollars. Fortunately for Japanese automotive manufacturers, their bankers are more understanding of these simple facts than American bankers.

Pipeline oscillations in inventory level, delivery time, and manufacturing investment are set up by what appears on the surface to be simple and easily avoidable circumstances. Consider one or two primitive examples of the strange effects of pipeline oscillations on a business [4]. If final demand for a product increases, the retailer may need a higher level of inventory to support the increase in sales and so on, back up the distribution chain.

Assume that retail sales of a particular product item increase 2 percent. The retailer might be expected to increase inventory by 2 percent to handle the increased sales level at the same level of effectiveness. Thus the distributor is presented with a 4 percent increase in demand. The distributor may tack on another 4 percent increase in inventory to continue filling orders at the same level of promptness as before, presenting the regional warehouse with a total 8 percent increase. In response, the manufacturer may project a need for a 16 percent increase in production rate.

A delay in response to changes in order rate along the distribution chain might appear to smooth matters, but this is not necessarily so. Even a temporary change in order rate can have a large impact. Suppose that four customers buy a ton of steel castings each week from one vendor. The vendor must produce 4 tons a week. Assume that the vendor has an order lead time of three weeks and thus has orders for 12 tons on the books at any one time. Suppose that the vendor receives an order for one extra ton in one week. Because this sales increase may be temporary, the vendor does not want to make the investment that would be required to increase the production rate; instead the vendor notifies regular customers that the order lead time has increased to four weeks. To be covered for this extra lead time, each regular customer must place an order for an additional ton. Thus the vendor's order backlog jumps to 17 tons, and the vendor must increase the lead time to five weeks, and the cycle continues. As Mather points out, "Amazingly, when inventories are at their highest, shortages are at their worst."

The reader unfamiliar with business may feel that these examples are frivolous, but remember we have placed ourselves in the position of knowing what is in each stakeholder's mind. Of course, this is not the way really things are. In actuality, the casting example is quite realistic. Mather reports that from 1974 to 1984 lead times for castings went from three weeks to 50 weeks and back again. Table 8.3 shows this oscillation for several other commodities.

TABLE 8.3 OSCILLATION FROM MINIMUM TO MAXIMUM ORDER LEAD TIME FOR VARIOUS COMMODITIES

Commodity	Order Lead Times	
	Minimum (weeks)	Maximum (weeks)
Castings	3	50
Resistors	4	45
Semiconductors	6	60
Bearings	8	108

Source: H. F. Mather, "The Case for Skimpy Inventories," *Harvard Business Review*, January-February 1984.

Order lead time is not the only variable affected by pipeline oscillations. Prices vary wildly, and investments in new plant capacity may be thrown out of kilter. In the spring of 1984 *Fortune* magazine pointed out that although demand was soaring for integrated electronic circuits and manufacturing capacity was increasing rapidly, it was clear that a "brutal shakeout" would occur as early as 1986 [5]. In reality, the recession hit the electronics industry only six to nine months later. In the recession of 1980 prices for certain memory chips dropped 70 percent in one year, in the face of falling demand. In 1984 prices had advanced 50 percent and were still climbing at the time of the *Fortune* article. Thus in spring 1984 incentives for manufacturing plant production capacity expansion looked ''irresistible''; not so by the summer of 1985, however.

The roller coaster didn't start for semiconductors in 1980. A sales buildup occurred in the early 1970s, followed by shortages, order delays, price increases, plant capacity increases, and a collapse in 1975, again followed by a buildup and another collapse in 1980. Late in 1988 Apple helped produce a shortage of chips and steep price rise in these vital components by stock piling. It then felt impelled to increase the Macintosh II sales price by 28 percent. Customer resistance resulted in slowed sales and a price cut for the Mac II in early 1989. The saga continues. If the various competitors in the semiconductor industry all have the same foreknowledge, why do they plunge ahead? Because in the race for market share, if one competitor cannot offer the same or shorter delivery time, customers go elsewhere, perhaps even overseas to Japan, as happened in previous downturns. So even foreknowledge is not sufficient to eliminate the problem.

8.5 STOCKLESS PRODUCTION: KANBAN

As revolutionary as the implications of reducing raw material and OEM parts inventory are, stockless production is even more disturbing to the typical American manufacturing plant supervisor. To put the WIP inventory in perspective, consider the situation in one of the most advanced high-tech plants in the United States. This advanced plant is in a firm universally acknowledged to be a world leader in every respect.

The production line concerned produces the read-write head for an advanced state-of-the-art computer disk drive memory device. The clearances are unimaginably tight, and quality control must approach perfection. The read-write head assembly line has 47 workstations, each separated by a WIP store. Including the incoming store and the stock of completed heads at the end of the line, there are 48 queues. The total inventory of heads at any one time in this production department averages 800 units. Thus an average of 753 pieces are idle at any one time in the 48 queues, an average of 16 idle pieces per queue. The ratio of active pieces being worked on to idle parts in the queue is 47 to 753, or 1 to 16.

The value added to the read-write head as it passes down the line rises rapidly, and near the end of the line each unit represents several thousands of dollars of inventory value. Make the conservative assumption that the average unit value while in the department is $1000. The idle WIP in this one department represents about $750,000. Efforts are under way to reduce the WIP to a ratio of two idle to one active work piece, which represents an asset savings of over $650,000.

The goal of stockless production is even more ambitious than reduction in WIP because it seeks totally to eliminate idle stock between machines in a department [6]. Stockless production is implemented first by moving all stock from stockrooms and warehouses to holding points adjacent to work centers. Next, stock is divided into containers of parts, each sufficient to make the standard number of units in a production batch. It is quite likely that at first the amount of idle stock may be too large to fit into holding points along the production line, and attention needs to be focused on reducing this stock size. Note also that the standard unit of production must be defined. Thereafter, it must be maintained despite pressure to produce odd lot sizes on a rush basis. Seek to achieve a steady flow of standard lots.

Once holding stocks are reduced to manageable levels, attention can be focused on streamlining the work flow. Nakane and Hall point out that stockless production is a pull system of production control that requires two kinds of control cards, or tokens, a move token and a production token. Japanese manufacturers call these *kanban* tickets, and the process is sometimes given this name (see Fig. 8.1).

The move token authorizes transfer of one standard container from the outbound stock point of one workstation to the inbound stockpoint of the next work center. This move token is obtained down the line from a container that has been authorized for production. In Fig. 8.1, note container 1 at the outbound stock point. Nothing can move into or out of this workstation, and no work can be produced by it until authorized by a kanban token. As the process begins, container 1 does not have a move token, so it sits. Examine the input side of the workstation; because no produce token exists, the station is not working. All kanban tickets on the input side are move tickets not produce tickets. Completed work cannot leave the outbound point either, because no move tickets are available. This factory is not yet at peak efficiency because two containers are at the inbound point and two at the outbound point. As soon as possible some tickets should be destroyed, forcing a shrinkage of the amount of WIP on the floor.

The regulatory process proceeds as follows. A move card is brought to the station from down the line and is placed on container 1. Container 1 can now move down the line, and the produce card formerly attached to container 1 is freed. The produce card, or token, authorizes completion of work on the standard container, labeled 2 in Fig. 8.1, to replace the container just taken from the outbound stockpoint of that workstation. When

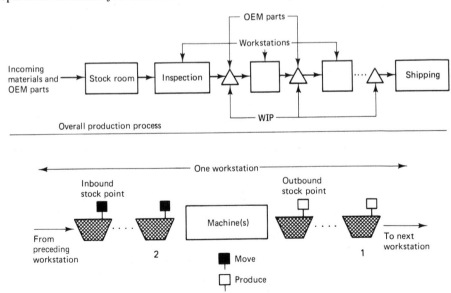

Figure 8.1 Detail of a stockless production process.

the produce token is placed on container 2, this action frees the move card attached to it, and this move card is taken to the outbound stock point of the preceding, upline workstation, continuing the process. The rules are simple but strict.

1. Always use standard containers filled with the correct number of parts.
2. Never move a standard container forward to the next workstation without authorization by an unattached move card.
3. Never produce a standard container of parts without the authorization of an unattached produce card.

Because only one card exists per container, WIP inventory is limited to the number of cards issued. Note that the containers themselves can serve as tokens. Initially, when extra stock is moved out of the warehouse onto the production line, extra tokens must be issued, but the drive to reduce the number of extra tokens must be unrelenting. Production holdups, extra cards, and delays signal deficiencies to be corrected in the production process rather than deficiencies in the control system.

The standard production lot size in the kanban system is generally smaller than the so-called economic lot size in conventional American production management. This factor and others result in difficulties with stockless production when introduced in American factories. It is necessary to gain workers' cooperation if stockless production is to succeed. Concepts such as rush orders and red-ticket production of units with unavailable parts omitted must be abandoned. Even in Japan, say Nakane and Hall, it can take up to five years to implement stockless production, but the results are worthwhile.

The philosophy that motivated the development of kanban and stockless production is quite different from traditional American production management philosophy. Traditionally, American producers strive to keep the line moving. Even when American workers are allowed to stop the assembly line when they detect production faults, they do not exercise that right. The Pontiac Fiero line personnel were encouraged to stop the line if they detected faults. Yet in the first six months of production not a single voluntary employee line stoppage was initiated.

Red-ticket production is part of the traditional American approach. It is a strategic error disguised as a tactical victory. Often in American manufacturing practice one or more parts or subassemblies will run out, become ''stock-out,'' during a production run. The part or subassembly is not available when needed in the assembly process. Conventional American practice has been to continue operating the line if possible, marking the incomplete unit with a red tag describing its deficiencies. The red-tagged item is held for rework at the line's end until the missing item appears.

8.6 FINISHED GOODS INVENTORY

One of the curiosities of pre-Iacocca Chrysler was its so-called sales bank. All over Detroit vacant lots were filled with unsold Chrysler cars, produced without orders and with no delivery destination. Although it has long been standard auto practice to record an auto as sold when it rolls out of the factory on its way to a dealer, it was also universal practice not to produce units without a dealer destination. The old Chrysler contravened this standard practice when, in desperation, it conceived the notion of producing cars, recording them as sold, and storing them on vacant lots. The practice created a semblance of good health on paper but was the final gasp before creditors discovered the truth.

The huge parking lot at the Michigan State Fairgrounds on 8 Mile Road in Detroit was once covered with a Chrysler sales bank. Some of the unsold autos had been exposed to the elements for a full year, and wind had blown dirt and water through door and window gaps. Tufts of grass grew on the floor and seats of many of these sad excuses for "new" cars.

The public relations notion of labeling unsold, unwanted products as a sales bank may be laughable now, but many other firms produce for inventory. What is that but a sales bank? Why do they do this? It seems to be a way of smoothing the manufacturing and distribution process, but isn't it just another cosmetic coverup that attempts to avoid reality? A sales organization has to accept the cost of carrying a certain level of finished inventory to service typical order levels, but it is another matter entirely for an organization to lose control of its production schedule.

Finished goods (FG) inventory is not totally bad. In fact, studies show that to purchasing agents, a supplier's available FG inventory is the single most important item influencing the buying decision [1]. Thus manufacturers must exercise careful judgment in balancing customer desire for quick response and the financial health of the firm. Think of a curve of FG inventory availability and presumed customer satisfaction versus the inventory investment, as shown in Fig. 8.2. A 100 percent availability of all FG product items at all times can be achieved only at an unacceptable cost. Thus, in practice, the organization must consider the benefit-cost ratio of moving between points A and B. In reality, however, firms typically operate at point C because the supply of all items is not in perfect balance with demand. Point C represents a condition of excess of one group of products and insufficient inventory of other products. With better management of the inventory mix of products, a company can move to better customer satisfaction at the same cost (point B) or reduce costs without reducing current customer perceptions (point A). The curve represents a perfectly balanced inventory mix of various products to provide the desired product availability.

Next is presented a simple graphical approach to understanding and controlling the finished goods inventory of a make-to-stock manufacturing enterprise suggested by Armstrong [1]. This graphical approach provides an insight to the problem without becoming bogged down in mathematical complexity. The approach encourages comparison of inventory and production performance by inventory sector, by profit contribution, and by flagging exceptions.

We are familiar with two measures of the inventory situation, namely the total inventory investment shown as an asset on the corporate balance sheet and the concept of annual inventory turnover. These indicators are too gross to be of value in managing the FG inventory, however. Two other simple measures, fill rate and utilization rate, are

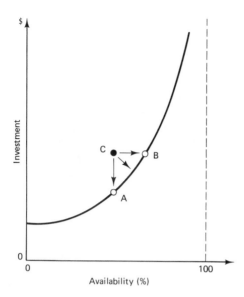

Figure 8.2 FG inventory investment versus FG inventory availability.

easy to determine and are effective control signals. By defining the terms in percentages rather than using absolute values, it is possible to make simple and effective comparisons across product lines.

The fill rate is defined as the percent of product orders that can be filled out of FG inventory. The ideal fill rate need not be 100 percent. Suppose that Ajax defines its target fill rate as the range between 60 percent and 80 percent. This means that Ajax wants to be able to fill, pack, and ship between 60 percent and 80 percent of orders for a given product within the standard inventory control period. Four to six weeks is a commonly used period in manufacturing.

Utilization rate is the proportion of a product manufactured that is shipped in the inventory control period. As with fill rate, the ideal utilization rate need not be 100 percent. A low utilization rate indicates that inventory is available as a safety stock or buffer. It indicates an idle corporate asset. Ajax might decide on a target range of between 40 percent and 60 percent for each product line, simply for example. Each parameter is now defined more carefully.

Total available (FG) inventory   Measured for an overall product line or by specific product items in number of units.

Current demand   Measured for each product line in number of units ordered for delivery during the current month (four weeks). After the middle of the month, add demand for the next month (for a maximum of six weeks total).

Production requirement   Defined as the current demand in units by product item within each product line that available FG inventory cannot satisfy.

Quantity fill   Current demand less production requirement.

Percent fill   Quantity fill divided by current demand.

Percent utilization   Quantity fill divided by total FG inventory.

For organizations with a small number of high-ticket products, all calculations can be done on a specific product item basis. However, such a detailed procedure may be prohibitively expensive for firms producing many small-ticket items. In this case a company can present data by product line. A mixed approach is also possible. Current industry practice will dictate modifications in desired order backlogs, delivery times, and measurement intervals. A. P. Sloan, Jr., imposed a ten-day reporting period on an industry accustomed to a casual once-a-year eyeball estimate. Many precise reporting and control details are best tackled by quality circles.

The bar graph in Fig. 8.3 can be set up for each product line (or item) and plotted for each control interval. In the "good relative balance" case in the upper left, Ajax can fill demand and is using its inventory promptly. In the "low inventory" case in the upper right, Ajax utilizes its inventory well but is failing to fill promptly. In the "low demand" case in the lower left, Ajax fills orders but has a slow turnover, and in the "bad balance" case in the lower right, Ajax is mismatching demand and current inventory.

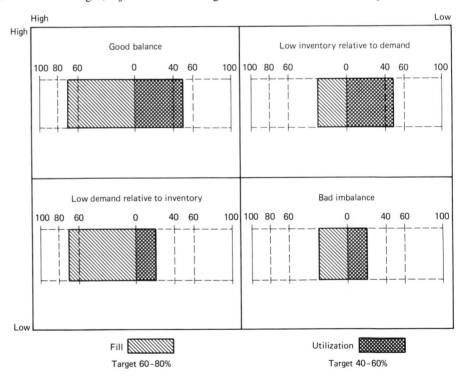

Figure 8.3   Fill rate, utilization rate, and FG inventory balance targets at Ajax. (After Armstrong [1].)

Figure 8.4 uses the graphs of Fig. 8.3 to represent what Armstrong calls a snapshot of the inventory across product lines at one instant in time. Expressing fill and utilization as percentages allows the company to see at a glance the inventory position over the whole product line and exposes off-center conditions. A firm could also use this portrayal for one product or product line at given time intervals, a sort of moving picture.

Armstrong suggests another moving picture to be used for total FG inventory control or for a single product line, shown in Fig. 8.5. As the ten-week period begins, a significant production backlog exists to meet current demand. This backlog is worked off as the weeks go on. At the tenth week inventory exists to meet future demand. Utilization is slipping in this period, and fill is riding at a high level. This situation bears watching. Can the enterprise safely reduce inventory investment on the item? The answer depends on its ABC class. Although it is only a ten-week history, a chart of this nature covering a longer period permits the manager to analyze trends and balance patterns.

Ajax should probably not use the same target ranges and fill rate and utilization rate for all its products. The standard adopted should be based on an evaluation involving ROI. Armstrong's calculation method is the "turn and earn" process and the gross margin return on inventory investment (GMROII).

$$\text{GMROII} = \frac{\text{``Turn'' sales (\$)}}{\text{average inventory at cost}} \times \frac{\text{``Earn'' gross profit (\$)}}{\text{sales (\$)}} \qquad (8.1)$$

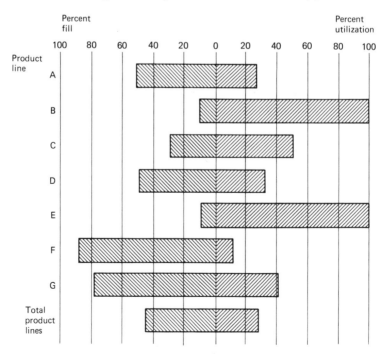

Figure 8.4 Snapshot of fill rate and utilization rate by product line. (After Armstrong [1].)

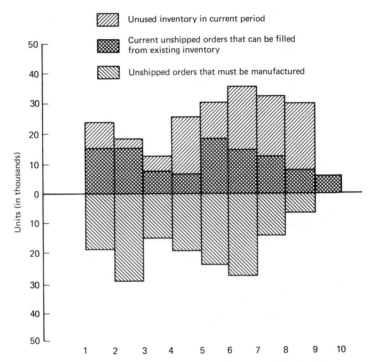

Figure 8.5 Weekly FG inventory on an absolute basis for one product line that covers eight top-selling items in the enterprise. (After Armstrong [1].)

Unfortunately, neither of the terms in Eq. (8.1) matches the definition of terms already defined [7]. The "earn" term is gross profit divided by sales at cost and is thus a ratio. It is not gross margin as defined in Eq. (5.1). The "turn" term is not the inventory turnover rate defined in other chapters. Inventory turnover rate is defined here as sales at cost divided by average inventory at cost. Thus

$$\text{Turn} = \frac{\text{turnover rate}}{1 - \text{GPM}} \qquad (8.2)$$

$$\text{GMROII} = \text{GPM} \times \frac{\text{turnover rate}}{1 - \text{GPM}} \qquad (8.3)$$

Suppose for its metal waste basket line Ajax has a gross profit margin (GPM), of 41 percent and an annual inventory turnover rate of 4.9 times. Then

$$\text{GMROII} = 0.41 \times \frac{4.9}{1 - 0.41} = \$3.41 \qquad (8.4)$$

For each dollar invested in inventory in this product line, Ajax grosses $3.41. Continue now to develop a monetary measure for setting the target range for fill and utilization. Pareto observed empirically that an 80-20 ratio is ubiquitous in human affairs. Typically speaking, 20 percent of customers account for 80 percent of sales, 20 percent of the product line accounts for 80 percent of the sales volume, and 20 percent of one product line accounts for 80 percent of the inventory. Unfortunately, it is not the same 20 and 80 in each situation. The point then is to focus on the 20 percent of the inventory that produces 80 percent of sales. Product items should be segmented and appropriate fill and utilization ranges established for each.

A different empirical concept based on the same phenomenon is the ABC rule. Rather than the two-part division suggested by the 80-20 ratio, the ABC rule commonly used in practice makes a three-part division. It starts by listing all product items in descending order with the highest dollar sales items at the top. A items are those that account for 50 percent of total sales. C items are the bottom 50 percent of items by count (not sales). The B items are those remaining [8]. Other ABC division rules also exist in practice.

Making this segmentation on product items can provide surprises. Not all of the A items are obvious at first. Often A items constitute a very small group, and a higher portion of the FG inventory investment should be concentrated on these items to ensure product availability and increased profits. On the other hand, C items move slowly and represent a drag on profits. Some C items may actually represent a loss to the firm. Availability on C items should be reduced even to the point of risking an occasional stockout. Taken together, shift of FG inventory investment out of C items into A items can increase profits and reduce the total FG inventory investment.

8.7 MANAGING THE INVENTORY INVESTMENT

Inventory has three main components: raw materials (RM) and OEM parts, work in process (WIP), and finished goods (FG). The value of the inventory consists of more than materials. It includes labor costs and manufacturing overhead. Obviously, attention should be concentrated on identifying and then reducing the most expensive components of the inventory investment, so that for a given sales level ROI is maximized.

Davis suggests a simple graphical approach to the problem [9]. The dimensions of the axes of the chart in Fig. 8.6 are so chosen that an area in the diagram represents the value of the inventory component being portrayed. Thus a large area signals a larger inventory investment than a small one. The cost-conscious manager concentrates on reducing the larger areas in the diagram first.

Take Ajax Metal Products Corporation as an example of this graphical representation. The Ajax balance sheet in Table 5.3 lists 1980 total corporate inventory assets at $8,100,000. Although it is better to break this figure down on a divisional basis, we shall save that for an exercise. Here we will deal with the problem on an overall corporate basis. The BS does not break down total inventory into its three main categories, and this division is needed to proceed (see Table 8.4).

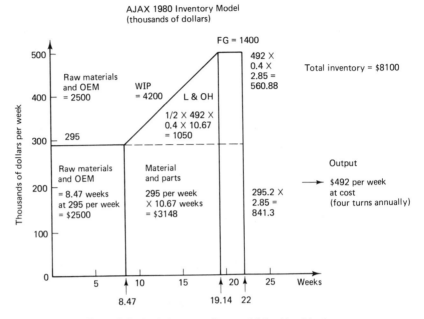

Figure 8.6 Davis inventory flow model for Ajax Metals.

We will make certain simplifying assumptions here, but the reader will see from the example how more detailed information could be employed if desired. For example, we will assume average dwell times for raw materials and finished goods and that value is added in the production process in a linear fashion.

TABLE 8.4 DETAILED BREAKDOWN
OF CORPORATE INVENTORY

Ajax Metals: Corporate Inventory Components for 1980 (in thousands of $)	
Raw materials and OEM parts	2,500
Work in process (WIP)	4,200
Finished goods (FG)	1,400
Total	8,100

Note from the Ajax IS for 1980 (Table 5.6) that the cost of goods sold is $24,600 (in thousands). Assuming a constant production level over a 50 week year yields a weekly shipping level, at cost, of

$$\frac{\$24,600}{50} = \$492 \frac{COGS}{week} \tag{8.5}$$

Given the value of the FG inventory from Table 8.4,

$$\frac{\$1,400}{\$492 \text{ per week}} = 2.85 \text{ weeks of FG stock} \tag{8.6}$$

To proceed further requires an additional assumption concerning the ratio of materials value in inventory to the value of labor and overhead. From the data for Ajax, San Antonio, in Table 6.4, note that RM & OEM represent approximately 60 percent of COGS, while variable labor and variable OH is covered in the remaining 40 percent. In the absence of firmer data, take these figures as an approximation for the corporate-wide numbers.

Taking the RM & OEM inventory of $2,500 and dividing by the parts share of the COGS yields the weeks of stock.

$$\frac{\$2,500}{\$492 \times 0.60} = 8.47 \text{ weeks of RM \& OEM stock} \tag{8.7}$$

The value of WIP consists of the value of material and parts plus the value of labor and manufacturing overhead. Thus

Materials contribution + labor & OH = WIP inventory balance content

$$(\$492 \times 0.6) \text{ (number of weeks)} + \frac{1}{2} (\$492 \times 0.4) \text{ (number of weeks)} =$$

WIP inventory balance

Let n be the number of weeks. Employing the relation for the area of a triangle,

$$(\$295.20) \times (n \text{ weeks}) + 0.5 (\$196.80) \times (n \text{ weeks}) = \$4,200$$

Solving for n yields

$$n = 10.67 \text{ weeks of WIP stock} \tag{8.8}$$

These data can now be plotted on the Davis diagram in Fig. 8.6. Ajax Metals at the corporate level shows a high proportion of inventory investment devoted to incoming materials and the materials component of WIP. If the same balance holds true at the divisional level and for individual product lines, attention to kanban and stockless production methods should be given higher management priority initially than labor productivity studies and final goods distribution.

Ajax need not give much concern initially to the FG stock because the chart area, and thus the dollar value, is small when compared with other components. Note that the labor and OH component of WIP is not large on a comparative basis. However, the materials component of WIP and the total incoming materials value represent by far the largest portion of the total Ajax inventory investment. Focusing on improving labor productivity and handling of finished goods shipments can wait until just-in-time inventory control and stockless (kanban) production methods have been installed.

EXERCISES

1. In spring 1984 Chrysler reported inventories at $1.3 billion [10]. Several years earlier this figure stood at $2.1 billion. At the current prime rate, what annual interest savings does this corporate inventory reduction represent?

2. When a stockless production system is first started, all factory inventory is broken down into standard production containers and given kanban tags. The aim is gradually to reduce this inventory by eliminating the extra containers. Exactly how would you suggest that this goal be accomplished?

3. Note that JIT inventory control (Sec. 8.3) implies close coordination with suppliers. Check the literature on this issue and discuss.

4. In Eq. (5.4) in Chapter 5, the average annual inventory turnover for Ajax Metal Products was calculated as 4.07 turns per year, identified as a convenient approximation. Some American manufacturers make additional approximations in this calculation. For example, one East Coast manufacturer calculates inventory turns using only the value of RM and OEM in Eq. (5.4). Suppose that Ajax made this assumption. Comment on the impact of this departure from accepted practice.

5. Develop the Davis flow model for model A at Ajax, San Antonio. Assume 50 weeks per year, RM and OEM at 70 percent of COGS, and OH at 30 percent of COGS. Use Table 6.2 and the following information.

<table>
<tr><td colspan="2">Corporate Inventory Components (in thousands)</td></tr>
<tr><td>RM and OEM</td><td>$ 70,000</td></tr>
<tr><td>WIP</td><td>95,000</td></tr>
<tr><td>FG</td><td>23,400</td></tr>
<tr><td>Total inventory</td><td>$188,400</td></tr>
</table>

6. Garrett Hardin's concept of the "tragedy of the commons" shows why the logical behavior of individuals under certain circumstances is destructive for the group as a whole, even when the facts are known to all parties [11]. Is the buildup and collapse of inventories described in Sec. 8.4 an example of the same kind of behavior? Explain.

7. Update Table 8.2 with data for 1985 and also current data. Comment on the implications of the trends.

REFERENCES

1. D. J. Armstrong, "Sharpening Inventory Management," *Harvard Business Review*, November-December 1985, pp. 42–58.

2. J. E. Gibson, "Henry Ford: The First Japanese-Style Manager", in *Essays on Management* (in press).

3. *Forbes*, August 1, 1983, p. 122.

4. H. F. Mather, ''The Case for Skimpy Inventories,'' *Harvard Business Review*, January-February 1984, pp. 40–42.

5. B. Uttal, ''The Coming Glut of Semiconductors,'' *Fortune*, March 19, 1984, pp. 125–130.

6. J. Nakane and R. W. Hall, ''Management Specs for Stockless Production,'' *Harvard Business Review*, May-June 1983, pp. 84–91.

7. The variations in definitions used for business parameters can be annoying. However, this variation is a fact of life in current business practice. Always take great care in defining terminology.

8. There are other ABC division rules; for example, see J. L. Colley, Jr., R. D. Landel, and R. R. Fair, *Operations Planning and Control* (San Francisco: Holden-Day, 1978), p. 76.

9. E. W. Davis, *Inventory Flow Models* Publ. No. UVA-OM-120 (Charlottesville: University of Virginia, Darden Graduate School of Business Sponsors, 1980).

10. *Wall Street Journal*, March 23, 1984, p. 8.

11. G. Hardin, ''The Tragedy of the Commons,'' *Science*, December 13, 1968, pp. 1243–1248.

CHAPTER

9

Manufacturing Production Control

9.1 INTRODUCTION

> Manufacturing engineering has had a blue-collar, greasy fingernail, low-paying, dead-end job image in engineering schools since World War II, and all that has to change if you are serious about implementing the concept of integrated manufacturing systems in your company.

This comment was made by a university engineering department chairman at a conference in late 1982 called by a large computer manufacturer worried about difficulty achieving manufacturing quality and productivity in its plants sufficient to surpass foreign competition. The remark met with approval from not only academics in the audience but also the plant managers and other corporate executives among the hosts. This firm is one of the most advanced in the world and yet it has to change the way it manufactures its products if it is to continue to compete successfully in the international marketplace.

Since the late 1950s U.S. manufacturing engineering theory has progressed down a path now recognized as a dead end by leading American manufacturers. This recognition is universal in the high-technology areas of electronics and computers and is also dawning in the industries of smokestack America, such as steel and autos.

Manufacturing engineers focused on optimizing bits and pieces of the manufacturing process, but almost totally neglecting the overall view. Theorists sought to isolate those elements in the process susceptible to analysis by mathematical techniques such as queuing theory and linear programming while ignoring the process as a whole.

Manufacturing engineers in the United States have concentrated on eliminating bottlenecks in the production process. The implicit assumption buried in this approach is that process optimization consists of gradual, smooth adjustment of parameters of production and elimination of small glitches. But to make real progress, the basics of the process must be questioned. Just-in-time inventory control is one example of a "revolutionary" concept that has overturned micro doctrines such as economic production lot size and economic order quantity and threatens even the still-developing theory of material requirements planning (MRP). Kanban would never have been developed from the gradual evolution of micro doctrines.

JIT inventory control seems revolutionary because it is contrary to recent past American practice. Yet Lindgren pointed out in Chapter 8 that this concept was common to older American practice. U.S. manufacturing has been moving away from the overall optimum while suboptimizing a few elements of the overall process.

Perhaps without realizing it, manufacturing theorists in the United States have continued to follow the philosophy of Frederick W. Taylor and his nineteenth-century concepts of scientific management. Taylor believed in breaking manufacturing down to its elemental parts and optimizing each separate element in turn. He also developed an authoritarian concept of labor management, so-called Theory X, discussed in Chapter 10. It is interesting to note that Japanese managers, while basing their manufacturing approach on past American practice, have not allowed themselves to be trapped into unexamined premises such as the Taylor syndrome of suboptimization.

The optimization of the elemental parts of each worker's task was important at the time Taylor developed it, but by exclusive emphasis on this microscopic approach in the recent past, manufacturing engineers in the United States have neglected the connectiveness of the overall process. Japanese engineers did not make this error. We might call this nearsighted focus on the details of production suboptimization, while assuming that the overall process is correct, "manufacturing myopia." Section 9.2 gives a number of examples of this sickness.

9.2 MANUFACTURING MYOPIA

Examples of manufacturing myopia abound, and no industry segment has a monopoly. Do not assume that myopia necessarily consists of ignoring modern tools. It also develops through fixation on tools and techniques while ignoring the overall connectiveness of the manufacturing process. The General Motors Technical Center in the early 1980s took great pride in a huge computer simulation of the complete auto assembly production process. The actions and timing of each worker's effort were simulated, and manufacturing theorists spent months reassigning bits of tasks among workers to balance the line to ensure that the nominal timing of each worker's assignment took the same number of seconds. Meanwhile, GM ignored worker dissatisfaction that resulted in the chaos at the Tarrytown assembly plant and wildcat interruptions at Lordstown, and the company canceled an experiment in "stall building" despite evidence of growing worker pride in this experiment at one Detroit plant. The situation represented the Taylor syndrome with a vengeance.

Here are some additional examples of functional fragmentation in an integrated manufacturing system that has resulted in lost effectiveness.

Product Development and Design Separate from Manufacturing

Chapter 3 discusses RCA's breakdown between marketing and product development in the case of its VideoDisc concept. That illustration can emphasize the chasm between product development and manufacturing as well. The VideoDisc was developed at RCA's laboratories in Princeton, New Jersey, and was manufactured in Indianapolis, Indiana. The geographic distance is great, but the psychic distance between the two groups was even greater. With only minor poetic license, each group claimed to be barred from eating lunch in the other's cafeteria when on business trips to the other site.

Economic production of high-performance products requires excellent quality control, but the products must also be initially designed for manufacturability. The manufacturing engineer must forget the boast of the past, "I can build anything that design can put in an engineering drawing," and the designer must forget the conceptual wall that has traditionally separated the design room from the factory floor.

Inventory Purchasing Separate from Manufacturing

Chapter 8 discusses the importance of carefully integrating continuous, smooth manufacturing and the JIT approach to inventory control.

Marketing Separate from Manufacturing

Chapter 3 emphasizes the concept of market pull in contrast to technology push. But in addition to ensuring market acceptability for a new product, the marketing effort and production flow must be integrated. A number of high-tech companies went bankrupt because they encouraged a market growth that could not be met with a carefully manufactured product. Quality control suffered, or cash flow was insufficient to meet immediate needs, or both. Quality control is discussed in Chapter 10 and cash flow in Chapter 11. Osborne Computers is one high-tech firm that went bankrupt because its marketing people announced a computer its manufacturing people were not geared up to produce.

Personnel Management Separate from Manufacturing

The manufacturing engineer who ignores the human factor on the production line has lost sight of the main target. People, according to Taylor, are like other tools: the last thing a manager wants them to do is think. But perpetuation of this ancient aberration is responsible more than any other single factor for the problems American manufacturing faces. Chapter 10 cites a manufacturing firm in the telecommunications marketplace that successfully installed quality circles and saved over $1 million in the first year by listening to its employees. But only one more year later a hard-bitten Theory X manufacturing manager had killed the QC program.

9.3 THE INTEGRATED MANUFACTURING SYSTEM

In contrast to optimizing separate elements of the manufacturing process, as advocated by Taylor and followed today by his unwitting disciples, the newer concept of the integrated manufacturing system (IMS) focuses on overall product performance from initial design and anticipated market need through manufacturing and after-purchase maintenance.

As usual, "the grass is always greener . . . ," so the manufacturing engineer in an established, slow-growth industry envies counterparts in rapidly growing high-tech industries their availability of capital, while the high-growth company manager complains of chaotic competition and lack of understanding of marketplace needs. It is somewhat easier to implement IMS in high-growth sectors, so consider the more difficult cases of steel and autos, neither of which is a growth industry in the United States.

The American steel industry has had difficulties for several decades and has thus far been unable to deal effectively with its problems. Conventional wisdom within the steel industry says that it is permanently burdened with overcapacity, unacceptably high wage rates, and unfair foreign competition. An outside view states that Big Steel has ignored the need to improve productivity and cut costs through investing in new technology and has relied on the protection of tariff barriers and government intervention rather than meet competitors in the international marketplace. Even in 1989 when Big Steel showed an operating profit, demands for government steel import quotas continued.

Recall that U.S. Steel was originally formed in 1901 by J. P. Morgan as a cartel with the stated purpose of buying out Andrew Carnegie and raising the price of steel. Carnegie was a fierce competitor in the relatively new, high-growth steel business of the day and drove hard to reduce costs and increase market share. The older, more conservative steel companies could not compete successfully with him in the marketplace.

Despite production levels well below 70 percent of capacity, threatened bankruptcy by Weirton Steel, and huge losses by other major steel makers in the early 1980s, new steel mills were constructed in the United States and operated at a profit. How can this be? How could upstart "minimills" know more about steel making than Big Steel? The answer lies in small, specialized, highly automated mills located close to their markets and operated by well-paid, productive workers. The arguments as to why Big Steel couldn't do as well or even better don't make sense.

Equally specious seem U.S. auto companies' arguments as to why they continue to rely on the threat of tariff barriers, local-content laws, and voluntary restraint by Japanese manufacturers rather than competing in the marketplace. These restraints on trade cost American auto buyers over $1000 per vehicle. What will happen now that Japanese auto firms are building new plants in the United States and operating them with American workers?

9.4 LOCATING THE MANUFACTURING PLANT

Location theory is in a fragmentary state at best. Classical economic geographers such as Weber [1], Christaller [2], and Hotelling [3], have addressed the problem of why industrial enterprises locate as they do and whether location affects the economic efficiency of the firm, but spatial economics [4] has never been a mainstream activity of the economics profession. Several firms specialize in helping companies locate new factories, chief among them the Fantus organization, but Fantus relies more on statistics and experience than on location theory in making recommendations. It is a fact that more than half of all manufacturing plant location decisions are based on no data at all beyond where the CEO thinks would be a nice place to live.

Location factors that contribute to the economic efficiency of a firm are simple to list.

Transportation of Raw Materials. A steel mill should be located near its supply of iron ore and coal. *Near* in this case is an economic term. Iron ore can move a long way cheaply by boat, and thus most basic steel mills in the United States are on the coast, the Great Lakes, or major rivers. Modern minimills use electric furnaces to melt scrap and thus can be located near urban areas. When the cost of transporting raw materials is a major differential factor, the industry is sometimes labeled "raw material intensive."

Transportation of Finished Goods. Some industries such as bakeries are distribution intensive and should be located near their final markets.

Energy Intensive. The reduction of aluminum requires vast amounts of electricity, so aluminum smelters are located near sources of cheap electric power, usually government-subsidized hydroelectric plants.

Labor Intensive. The textile industry and the furniture industry need a source of reliable, low-cost, relatively unskilled labor. As labor costs rose in the Northeast and the Midwest earlier in this century, textiles and furniture manufacturers migrated to the southeastern United States. Consider the strategic implications of this migration by comparing Japanese industrial policy developed by MITI.

Tax Differentials. Fantus has found that, with the exception of a few industries, the determining factor in most location decisions should be the differential burden of taxation and local government subsidies.

Offshore manufacturing is a difficult subject to deal with objectively; yet ignoring the issue may lead an organization to make decisions that could result in corporate losses or even bankruptcy. Approach offshore manufacturing by means of the following example.

9.5 OFFSHORE MANUFACTURING AND THE PACKARD ELECTRIC CASE

The Packard Electric Division of General Motors, located in Warren, Ohio, has traditionally been the sole supplier of electrical wiring harnesses to the automotive divisions of GM. As some other GM divisions, Packard has a history of divisive, adversarial labor relations stretching back over several decades. The confrontational negotiating style of Local 717 of the International Union of Electronic Workers (IUE) has been reinforced by local plant management's macho self-image and take-it-or-leave-it bargaining style. The situation was exacerbated by Packard's position as sole supplier of vital electrical parts to the corporation [5].

Finally, in 1973, with a sense of failure perhaps, or possibly relief, Packard Division management adopted a policy of ''no new hires and no new bricks and mortar in Warren.'' In the next few years Packard opened two plants in Mississippi and three in Mexico. The average hourly wage in its Mexican plants is $1 an hour, while in Mississippi and associated subcontractor plants in the United States the average wage rate is $6 an hour. In Warren, Packard was required by its IUE labor contract to pay an average of $16 an hour, and for top assembly-line jobs to pay up to $19 an hour.

The situation in Ohio steadily deteriorated, and in the period from 1973 through 1977 Packard eliminated 2000 jobs in its Warren operations. In the same period the Warren-Youngstown area lost a total of 20,000 to 40,000 jobs through closings of steel mills and other plants. Chapter 10 considers labor-management style and techniques that might have helped at Packard, but this chapter concentrates on the question of economic efficiency.

Because of collective bargaining contract constraints, Packard was forced to pay unskilled labor in Warren at the same $16-per-hour rate as its more skilled and experienced people. Unskilled labor cannot be 16 times more productive than equally unskilled labor in Mexico, and by failing to deal with this fact, Local 717 forced GM to go offshore when its profits were threatened.

Once an industry is mature enough to have developed knowledgeable competition, it cannot survive in the open marketplace paying significantly higher wage rates without higher labor productivity. But there are other ways of improving labor productivity than hiring unskilled people at low wage rates. A new president of Local 717, working with new plant management, solved the problem at Packard; the matter is pursued in an exercise in Chapter 10.

Many U.S. manufacturers have moved overseas in search of even better wage breaks. But, as Markides and Berg in a carefully reasoned *Harvard Business Review* article point out, wages typically account for only about 15 percent of total product cost. Thus even if actual wage rates are halved, hidden costs of doing business overseas may more than eat up the apparent wage savings. In addition there are many strategic disadvantages to an overseas location, such as possible political instability and government takeovers, and difficulty in repatriating profits [6].

9.6 THE BUILD OR BUY DECISION

The decision of where to manufacture a product does not end with geographic location of the plant. Actually, the problem of where to locate a new plant is a rare decision, usually made at a high level in the corporate hierarchy. The manufacturing engineer is more commonly involved with questions concerning tooling to be used in a particular operation and whether the part should be produced in-house or purchased as an OEM part. Pros and cons for this choice follow.

BUILD IN-HOUSE

Complete Control of the Process. A manufacturing concern is in the business of manufacturing. Why should it farm out part of its work, *ceteris paribus*? Isn't it likely to lose control of delivery dates and quality control by farming out?

Profit Retention. An outside supplier has to earn a profit. Why doesn't the company keep that profit itself?

Proprietary Skills. A company should try to develop and improve proprietary skills in manufacturing its products. Why pay someone else to learn the trade?

BUY OUTSIDE

Lack of Skills. The company may not have the skills in-house needed to build a specialized part and may not use enough of the part to warrant building these skills.

Quick Turnaround. Often an outside specialized supplier can produce the order faster than the company can.

Higher Quality, Lower Cost. A specialized, high-volume producer may be able to supply a part not only more quickly but also more cheaply and of higher quality.

Smooth Use of Labor. An enterprise may wish to contract out production of a part to smooth its own labor utilization, even though it could make the part itself.

Access to Patented Technology. A better-quality part that is also cheaper to produce may be protected by patents.

Incommensurate Labor Skills. Even if a need for large quantities of a particular part is anticipated, it may be better not to produce the part in-house if the labor skills needed are at a significantly higher or lower level than those in the rest of the plant.

Ajax, San Antonio, chose to go outside to buy the combination locks for its four-drawer files, and in the venture analysis on the trash compactor, the company chose to go outside to buy the electric drive motor. Both of these decisions are consistent with the points on the list of pros and cons. But sometimes the decision can be rather subtle. Here is an example of incommensurate labor skills that forced a plant relocation decision.

A small electronics manufacturer in the Midwest, a relatively small division of a large conglomerate, made relays for use in the remote control of radios and other electronic gear. These relays are small and precise, with some models prepared especially to meet military specifications. The enterprise also made a line of cheap, heavy-duty telephone-type relays in a plant located in Kentucky, more than 100 miles from its main location in southern Indiana. The top manager of this firm was asked why he didn't consolidate the two plants, neither of which was very large. He said that in fact the two production lines had both originally been located in buildings in the same town, but the telephone relay line had to be moved to Kentucky.

When the assembly lines for the two products were located in the same town, the collective bargaining agreement required the seniority lists for the two operations be merged. Thus using standard seniority bumping procedure, workers could transfer from one plant to the other when work on a line went slack. This procedure was not successful, because when workers transferred off the telephone relay line to the electronic relay line, they were unable to meet the quality standards without a long adjustment period. The result was an unacceptably high spoilage rate and an inability to make the piece rate.

Transfers the other way also failed because workers from the electronic line worked to excessively tight standards on the telephone line and failed to make the piece rate. Of course, in time workers could learn to handle either line, but labor turnover was high, and much of it was part time so the labor force never reached equilibrium. The only solution that occurred to management was to locate the two plants sufficiently far apart to prevent workers from living between the two towns and commuting either way.

9.7 PLANNING THE PRODUCTION PROCESS

As is true for many other aspects of production management, plant layout is in the process of revolutionary change. If the discussion of JIT inventory control and stockless production is reviewed four distinct periods in plant layout can be discerned.

Job Shop

The earliest unit of shop organization was the workbench of the individual craftsman and that minimal organization was the form adopted in early factories, for example, the government arsenal at Harpers Ferry in the early nineteenth century [7]. Later, some organization was brought to the job shop by grouping machine tools by machine operation, separate from the foundry. This is the sort of shop that Taylor set out to improve at the turn of the century.

Assembly Line

Around 1913 Henry Ford and his associates developed the auto assembly line. Industrial engineers have spent the years since then optimizing the modified and improved assembly line, or flow process. Manufacturing operation texts focus on job shop and assembly-line optimization [8].

Important changes in the philosophy of design of flow, or assembly line, manufacturing have resulted from Japanese contributions since the 1970s—just-in-time inventory control and stockless production, for example. Both these innovations are management concepts. They do not require massive investments of capital; indeed, they reduce the need for plant capitalization.

Automated Workstations

Automation of the assembly line is taking place very gradually. First the movement of parts from one workstation to the next was automated by transfer machines. Then individual machine tools were placed under computer control using numerically controlled (NC) machine tools. Next tools were grouped around transfer robots in work cells, which is the current state of the art.

Team and Distributed Assembly

Soon it will be realized that the computer has rendered obsolete the entire concept of assembly-line mass production and permits manufacturing to move to the concept of distributed assembly.

9.8 FLOW-ASSEMBLY MANUFACTURING

Figure 9.1 is a diagram of a conventional flow manufacturing process. Note that this assembly process is not designed for just-in-time inventory control or stockless manufacturing and thus is an obsolete conception. The guiding principles in the design of this process follow.

- Maintain straight-through work flow. Bring the work piece and additional parts needed to the worker at the workstation.
- Keep individual work tasks simple and repetitive. Optimize individual worker assembly time and effort through time and motion study.
- Keep inspection separate from assembly work.
- Locate tool cribs and stockrooms near the work line.
- Use WIP inventory between workstations as buffers and to smooth minor perturbations in assembly operations.
- Encourage capital investment in specialized jigs, fixtures, and power tools to increase productivity.

Compared to the job shop environment, worker productivity is much higher in a well-designed assembly operation. Necessary work skills are significantly reduced, and in theory, quality control is easier.

Mass production through assembly line techniques has been responsible since its development by Ford for much of the increase in the standard of living in the United States in the past 80 years. Further, the free world relied on these techniques to produce

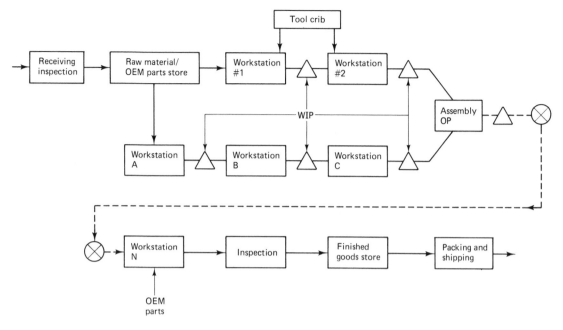

Figure 9.1   Schematic of a conventional flow manufacturing process assembly line.

the weapons needed in World War II. So careful consideration is necessary before the system is completely abandoned.

Modern Japanese manufacturers' developments in mass production are important and may or may not be considered revolutionary, depending on one's point of view, but they are logical and at least evolutionary. We will see in Chapter 10 that the Japanese have contributed more than inventory reduction and the rapid introduction of robots. Perhaps the most important contributions have been recognition of American manufacturing prophets such as Deming, adoption of statistical quality control and the realization that each worker has much to contribute to improving manufacturing operations through such devices as worker quality circles.

9.9 DISTRIBUTED MANUFACTURING

Perhaps the entire concept of mass production needs to be reevaluated, given the availability of totally new control tools, especially the small, reliable, and economical digital computer. Let us review the basic imperatives that formerly supported the concept of mass production, but which may no longer be valid.

In 1776 Adam Smith advocated the economies achievable by structuring a manufacturing process as a series of simple, repetitive steps performed by unskilled, and therefore cheap, labor. The same principle applies to the assembly line in the factory of the 20th century. To justify the continually rising wage rates of assembly line workers, it is necessary to increase productivity by placing them in charge of complex, expensive tools. Some tools on the typical Detroit auto assembly line cost hundreds of millions of dollars and can take three to five years to design and construct. Many thousands of autos must be constructed with these tools if the automobile is to be affordable by the masses.

Suppose it were possible to construct one of these expensive tools that would be flexible enough to be reconfigured not in years but in seconds and could be used to construct a variety of vehicles. Or suppose it were possible to make workers themselves so flexible that, working in small teams with simple and flexible tools, they could assemble the whole vehicle. The former option is feasible now, in principle at least, using computer-aided design and manufacturing (CAD-CAM) techniques and robotics, and the latter option is a technique called *stall building*.

Perhaps autos could be stall-built in individual auto dealer showrooms by the same individuals who maintain and repair the cars. Then the constraints of the huge, rigid, expensive factory in which thousands of individuals perform simple, boring, repetitive tasks could be eliminated. Obviously, distribution of parts would require fine tuning and computer control, but WIP and FG inventory would vanish if the customer were present to watch the car's assembly and drive it away immediately afterward. Just a thought starter.

9.10 OPERATIONS AND QUALITY CONTROL

The daily life of the manufacturing engineer or operations manager doesn't include much time for the topics in manufacturing textbooks. It consists of mundane tasks: finding somebody to fill in at a workstation because the regular operator had to take a child to the doctor, deciding whether to keep the line running without a needed part, and, of course, attending meetings with inexperienced staff people with wild ideas from the front office. I am familiar with one divisional plant that was ordered to install quality circles overnight because the corporate CEO in a distant city had seen a 20-minute segment on *60 Minutes* on TV the previous evening. Be that as it may, here are two typical textbook topics on manufacturing operations.

Balancing the Line

One of the legacies of Frederick Taylor's scientific management is the notion that workers should be tied to a single workstation and work brought to them. Although Taylor was concerned with optimizing work flow in a job shop, not an assembly line, the assembly line is in a sense the logical outcome of Taylorism. The need to balance the line is a direct consequence of the assumption that a worker can do only one job and in a totally prescribed way. In a series assembly process each worker's task must take precisely as long as each of the other workers' tasks, or WIP piles up at the workstation with the longest task, and workers ahead in the line go out for a break. The steps in balancing the line are as follows:

1. Break up the process into the simplest possible tasks.
2. Establish a time allocation needed to complete each task. The typical work cycle in U.S. auto assembly plants is 1 minute.
3. Establish precedence requirements for each task. Certain tasks must be completed before others can begin.
4. Assign tasks to workstations such that each station consumes approximately the same amount of time while obeying the precedence relations.
5. Arrange stations in series or in parallel.

For many years the process of line balancing was based on guesswork and experience. Recently, the major auto companies developed elaborate computer simulations to automate the process. But even the most careful time and motion studies establish only approximate times for each operation, and the process is deeply adversarial. The worker feels the need to take as much time for each operation as possible. To the worker, the industrial engineer with clipboard and stopwatch represents unfeeling authority. Discussions on matters such as the time allocation for tasks provides material for deeply felt conversations between union shop stewards and company supervisors.

Quality Control

By 1931 statistical quality control of industrial processes was standard practice in the United States [9], and quality control has long been a standard topic in college manufacturing engineering programs [10]. By the 1960s the theory was considered simple enough to present to freshmen engineering students [11]. Thus it is a puzzle as to why statistical quality control practices have not gained widespread use in U.S. industry.

When Japan began to rebuild its industrial base after World War II, Japanese managers resolved to seek out and adopt the best American industrial practices including statistical quality control (SQC). In the 1950s E. Edwards Deming was invited to give a series of lectures on SQC in Japan, and the topic became very popular. Deming became so widely respected in Japan as an authority on industrial quality control that MITI established an annual Deming competition for the best application of SQC and awarded gold medals and other prizes. In the 1970s one Japanese firm was so proud of winning the Deming gold medal that it hired a huge billboard near a Los Angeles freeway to celebrate this fact. Few Americans had heard of Deming or SQC, however. By 1980 the American auto industry had discovered SQC, and Deming had become a consultant to GM.

9.11 STATISTICAL QUALITY CONTROL

Statistical quality control consists, at its base, of the application of a few principles from elementary statistical theory. These principles can be applied to detect the presence of systematic errors in the production process, buried though such errors might be, beneath normal random variations. Once the existence of a systematic error has been detected in the SQC process, it is often a simple matter, in theory, to eliminate.

Shewhart [9] recognized that random variations of the mean value of small lots of a critical production parameter would themselves be distributed in a normal or Gaussian distribution about a mean value. The Gaussian distribution has the form of a bell-shaped curve and is particularly simple in that only two parameters, mu and sigma, are needed to define it. The mean or average value, the peak of the bell-shaped curve, is often designated by the Greek letter mu (μ). The Greek letter sigma (Σ) is often used to designate the standard deviation or spread of the variation about a mean value, that is, the width of the bell.

Figure 9.2 shows Shewhart quality control charts developed to display variations in a critical parameter of a production process. It is a characteristic of the normal distribution that 99.7 percent of all the values of the means of small samples in such a distribution fall within $\pm 3\ \Sigma$ of the mean, μ. Shewhart recommended that the critical dimension (the mean or expected value, μ) and the $\pm 3\ \Sigma$ limits be plotted on a chart. As long as less than 0.3 percent of the samples fall outside this range, the process is designated in control; that is, systematic variations are not present.

The normal distribution is plotted on the left of Fig. 9.2(a), and the right shows a process that is in control. Often, as here, the means of the small samples are shown as dots with a vertical line to the overall mean, μ. Figure 9.2(b) shows a process with a sudden jump, possibly caused by a new batch of raw material, and Fig. 9.2(c) shows a drift that is characteristic of tool wear. The wide scatter in Fig. 9.2(d) might be caused by an inexperienced operator or by a machine needing repair or replacement. The processes in (b), (c), and (d) are not in control.

Several sources of systematic errors in a production process exist, and it is important not to blame machine operators for systematic errors not under their control. On the contrary, machine operators' cooperation should be sought in detecting the presence and sources of systematic errors. Furthermore, random variations in the critical parameters of a production process must not be blamed on the operators, because the variations are caused by the process itself.

If random variations cause the process to produce units that fall outside the acceptable range, two choices exist. First, units within the acceptable range may be selected by final inspection and out-of-range units scrapped. Second, and more satisfactory in the long run, the process may be redesigned to meet the required tighter tolerances. This brief introduction to SQC is too abbreviated for functional application of the concept. It is included simply to call attention to its utility and importance.

9.12 THE AUTOMATIC FACTORY

Just as the early motorcar looked like a horse-drawn buggy or a motorized farm wagon, the early automatic factory looked like a conventional assembly line with robots instead of workers. The similarity represents not so much a lack of imagination on the part of manufacturing engineers as the desire to automate the factory gradually, replacing

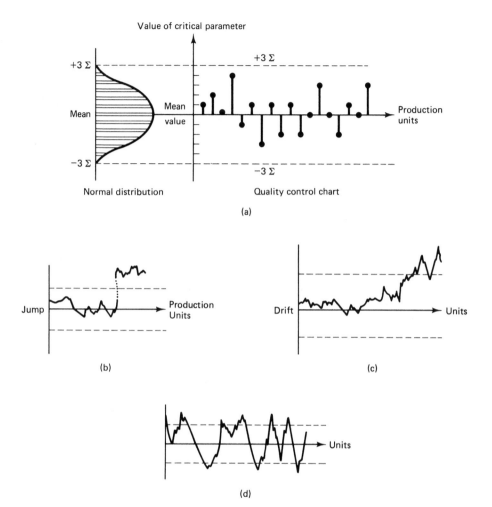

Figure 9.2   Shewhart quality control chart.

humans with robots one by one. First dirty and dangerous jobs such as spray painting and loading heat-treating furnaces were automated, and now a few factories are capable of operating completely without human operators. Fittingly enough, one of the first of these completely automated factories was the Fanuc robot manufacturing plant in Japan [12].

The totally automatic factory, that is, a plant without any human operators, is unlikely to prove economically viable. One suspects the Fanuc plant is more a master-stroke of marketing than a financially efficient venture. Experienced observers have remarked that automating the final 20 percent of a plant costs as much as automating the first 80 percent and should not be done. The paperless factory is a related idea and, in concept, it also seems within our grasp. The paperless factory is one totally under computer control, from original engineering design through final inspection, including all record keeping. But the 80-20 rule applies here as well. Perhaps the best way to describe the automatic factory or the paperless factory would be to describe a modern-day approximation.

Erie, Pennsylvania, is a traditional, heavy manufacturing city on the Great Lakes. Bucyrus-Erie, Hammermill Paper, a steel mill, and GE's diesel locomotive works are Erie's major employers. Industrial unemployment is high, and layoffs continue. Employment at GE was over 12,000, but had fallen below 7,000 in the fall of 1983. Business was bad in 1983, but even when it picked up again, most of the laid-off employees were not called back. The GE plant is large—more than 3 million square feet, and several dozen large buildings, some of which were built before 1914.

In 1979 GE started systems analysis of integrated automated manufacturing of diesel locomotives, continued in 1980 with an intense planning effort, and in 1981 received initial funding from corporate headquarters. The planning took a top-down, integrated manufacturing system (IMS) approach. It moved from analysis of business needs to overall technical concepts and further detailed breakdown, then, finally, to over 1000 major project definitions. Before completion of the seven-year project, GE will have committed over $500 million to PP&E, including construction of a completely new, "green field" diesel motor manufacturing facility in Grove City, 100 miles south of Erie, and over $250 million for engineering design. The expression "green-field plant" refers to a manufacturing facility built totally from scratch at a never-before-used site.

Although there were smaller GE test projects before, this one was the largest single automated flexible manufacturing project the corporation has ever approved. Since 1981 GE has announced projects of similar scope in lamp manufacturing, major home appliances, and aircraft engines. This project was approved while Jack Welsh was still vice-chairman, and his statements on market share are well known. Welch says that GE must be number one or two in a business sector or it gets out. In keeping with this policy, GE recently announced the closing of its Canadian steam turbine plant for lack of business. Steam turbines and electric generators are the heart of GE's traditional business, so imagine the shock produced across the organization by this decision.

GE expected a 25 percent reduction in direct labor and a 20 percent reduction in material cost from its Erie automation project. This reduction in material cost cannot come out of locomotive weight, which is needed for traction, but will come instead from reduced scrap and rework. In addition to traditional heavy metal cutting and fabrication, this project requires that the locomotive business operation develop a capability to design and produce electronic motor controls and locomotive-operator cab displays, among other items.

"GE, Erie, determines the specifications for the automation and tries to get someone outside to build it for us," explained Dwayne Shull, project general manager. "We design and manufacture locomotives, not machine tools. But sometimes it has been tough to get contractors to understand what we are driving at. We had Giddings and Lewis make the NC [numerically controlled] cutting tools, but we had to keep a lot of the work cell design and electronics project management in-house for lack of a qualified supplier" [13].

Direct labor has traditionally been the basis for evaluating efficiency in a manufacturing operation, but with the automated factory direct labor will be an increasingly smaller portion of the total cost of a manufactured product. Shull expected a more rapid manufacturing cycle and a 30 percent improvement in inventory turn. This efficiency provides a more flexible response to customer wants and permits tailoring of locomotives to specific needs, such as locomotive drawbar pull and wheel loading.

A sense of the magnitude of this project can be gained in a visit to a large room in Erie, its walls covered with system planning charts, tables covered with elaborate, three-dimensional models of flexible machining centers installed or under construction, and a complete 3D model layout of the Grove City plant under construction. Shull's characterization of the top-down IMS concept approach is supported by the diagrams and charts in this room. The complete approach extends over a seven-year period and covers over 1000 individual projects in the following six major manufacturing categories:

- Control
- Propulsion
- Diesel engine manufacturing
- Light metal fabrication
- Heavy metal fabrication
- Locomotive assembly and testing

Almost all of the 3.5 million square feet of manufacturing space in the existing Erie plant buildings will be rearranged. The whole project is on the computer: critical path method, project deadlines, dollars, resource planning, and scheduling. Shull reports, "We moved from the concept through block layouts via CAD and then to 3D layouts for the 11 separate buildings involved. Meanwhile, we are tracking technology on the outside so that we won't get blind-sided by something new. The aim here is the paperless factory from engineering design using computer graphics through manufacturing, quality control, test, and customer acceptance" [13].

One specific manufacturing operation to be installed at Grove City is the machining of the diesel engine cylinder jacket. The layout is straight through, and kanban production control will be used. Production cycle time will be reduced from three weeks to one eight-hour shift, and production personnel required will drop from 50 operators to three. The operators will do their own inspection, and full-color computer graphics displays along the line will report to the operators the current status of the production cycle including tool condition at each operation. A production report on current production rates and totals, as well as Shewhart quality control charts with 3Σ tolerances, or any other specs, can also be displayed. GE calls the system computer-assisted process planning (CAPP). Typical machining tolerances in this business are 1 or 2 mils.

At the operator's signal for a part number, an automated parts stacker finds and delivers the part to the workstation and then adjusts inventory control records. Computer-controlled, automated test cells run through the proper test cycle on completed engines and record all data. The test cell computer also times the various cell tests to produce optimum cogeneration of electricity and heat for the rest of the plant.

Because machine operations at Grove City are new and require more independent initiative from employees, most employees are newly trained recent hires, rather than employees transferred from the older Erie plant. One economy of the Grove City plant is the existence of only a half-dozen employee work classifications, job codes, while the Erie plant has over 50. Workers qualify for pay increments within a job code based on job performance and mastery of additional operations.

Automated cutting of steel plates is an operation retained at the Erie plant. The Erie operation is now the largest automated plate-burning mill in the world. Using interactive computer graphics, a manufacturing engineer lays out the parts to be cut from a steel plate, and the computer program positions the parts to minimize scrap. The computer transmits electronic signals to an automated crane that goes out to the storage yard to the proper stack of raw plates, picks one up and positions it underwater on the cutting table for the plasma torches. The computer even computes the operator's wages based on inches cut, number of parts cut, finish needed, and other tasks performed. All production workers at the Erie plant are on incentive pay. This whole burning mill operation is an example of the paperless factory concept in practice.

Electronics play an important role in modern locomotive engine control. Such parameters as engine speed optimization, diagnostics, graceful degradation of automatic controls, "fail soft," rather than catastrophic collapse can be placed under computer control. Computer displays for the driver indicate such items as failure diagnostics, engine cranking cycles, battery charging, fan control settings, and blower control settings. In automating the design and manufacture of the operation producing these electronic parts, two-thirds of all funds were spent on quality:

- Quality control
- Inspection
- Automated computerized testing

Some first-level automated operations are retained in this paperless factory. For example, in the cylinder liner machining operation a pick-and-place robot handles the 300-pound rough casting, from which 100 pounds or more of chips must be removed, by inside diameter (ID) grinding, outside diameter (OD) grinding, and spindle borers, all under numerical control. Each Unimate robot handler sits between two Ballard, NC machine tools. Two of these workstations have been in reliable operation for over four years. Previously, 12 to 14 operators were required for this operation. Two operators are used at Erie, and when the operation moves to Grove City, one will run both workstations. But GE does not plan totally operatorless production.

The most famous operation at GE, Erie, is the flexible machining system (FMS) for the main electric traction motor frames. When this operation was redesigned, 30 machine stations were converted to nine new, more complex stations. The old, full three-shift operation, which required 70 people, now uses only two direct labor people each for two shifts. At the same time a 10 percent increase in overall production has been achieved. Including direct labor, indirect labor, and supervision, GE estimates a 240 percent increase in this operation's productivity. Here are some additional statistics on this operation.

- Giddings and Lewis built the NC machine tools.
- Cycle time reduced from 16 days to one day.
- 24 percent less floor space used.
- 60 percent additional capacity achieved.
- Installation started in summer 1982.
- By summer of 1983 could handle production of six sizes of motor frames.
- By winter of 1983 projected to handle six more sizes.

In 1983 GE, Erie, was estimated to achieve a payback position on this project, primarily from reduction of direct labor costs, in approximately five years, assuming current production levels. If the business climate had improved, and GE had been able to utilize the 60 percent increase in plant capacity to increase sales, the payback period could have been reduced to approximately three years. Unfortunately, world demand for diesel-electric locomotives did not improve in the next four years, and GE did not realize the projected savings.

The most important fact to keep in mind in this example is the strategic decision process. By adopting plant automation, a company shifts variable direct labor costs to fixed costs in the form of PP&E investment. Achieving the estimated ROI and payback period requires that the estimated production level actually be maintained. Operating costs cannot be reduced through labor reductions if projected sales do not materialize. Furthermore, investment in factory automation requires that a company have a dominant market share in the industry segment to be automated. If not in this dominant low-cost-producer position and unable to make the capital investment required to gain that position, a manufacturer must consider leaving that line of business.

EXERCISES

1. Kenneth Boulding, a well-known American economist, has been quoted as saying, "I have discovered the real name of the Devil, it isn't Lucifer, it is sub-optimization." Does this statement capture the meaning of Sec. 9.1? Comment.

2. Section 9.3 suggests that stall building and computer control of inventory delivery might be economically feasible. Pick one specific difficulty with this concept and suggest a solution.

3. In Sec. 9.3 it is said that despite surplus capacity and bankruptcies in the American steel industry, some steel producers are expanding and earning profits. Check the business literature, *Forbes, Harvard Business Review*, and other publications for companies such as North

Star Steel, Chaparral Steel, Birmingham Steel, Co-Steel, and Nucor. Report on these steel minimill operators. Are they successful? If so, cite their success factors [14].

4. The automatic factory is discussed in Sec. 9.12. Near the end appears the remark that by adopting plant automation, a company shifts variable direct labor costs to fixed costs in the form of PP&E investment. Comment on payback period, profitability, and other implications of the decision to automate, given increases and decreases of 20 percent from pro forma production estimates.

5. What are the financial implications of substituting capital investment for labor investment in a product? What does this substitution imply about maintenance of market share? Should a company invest in a strategic business unit (SBU) to which it was not deeply committed? Relate to the strategic decision by GE in the locomotive business.

REFERENCES

1. A. Weber, *Theory of the Location of Industries* (Cambridge: Cambridge University Press, 1957).

2. W. Christaller, *Die Zentralen Orte in Suddeutschland*, trans., C. Baskin (Englewood Cliffs, N.J.: Prentice-Hall, Inc., 1966).

3. H. Hotelling, "Stability in Competition," *Economics Journal*, March 1929, pp. 41–57.

4. R. D. Dean, W. H. Leahy, and D. L. McKee, eds., *Spatial Economics* (New York: Free Press, 1970).

5. "The Revolutionary Wage Deal at GM's Packard Electric," *Business Week*, August 29, 1983, pp. 54–56.

6. C. C. Markides and N. Berg, "Manufacturing Offshore is Bad Business," *Harvard Business Review*, September-October 1988, pp. 113–120.

7. M. R. Smith, *Harpers Ferry Armory and the New Technology* (Ithaca, N.Y.: Cornell University Press, 1977).

8. E. S. Buffa, *Modern Production/Operations Management*, 6th ed. (New York: Wiley, 1980).

9. W. A. Shewhart, *Economic Control of Quality of Manufactured Products* (Princeton, N.J.: Van Nostrand Co., 1931).

10. A. V. Feigenbaum, *Total Quality Control*, 3rd ed. (New York: McGraw-Hill, 1983).

11. J. E. Gibson, *Introduction to Engineering Design* (New York: Holt, Rinehart and Winston, 1968), pp. 87–92.

12. F. L. Schodt, "In the Land of the Robots," *Business Month*, November 1988, pp. 67–75.

13. Personal communication, September 13, 1983.

14. "U.S. Minimills Launch a Full-Scale Attack," *Business Week*, June 13, 1988, pp. 100, 102.

CHAPTER

10

The Quality of Work Life

At Hewlett-Packard . . . we found that as much as 25 percent of our manufacturing assets were actually tied up in reacting to quality problems and we decided that through pursuing quality we could achieve lower production costs and improve our competitiveness. [1]

10.1 INTRODUCTION

Perhaps the most important element in managing the integrated manufacturing system is the management of manufacturing personnel. Yet until recently this issue was totally removed from the central concern of manufacturing managers and relegated to a minor role. All too often in the American manufacturing canon ''personnel management'' as a job category was thought to be the Siberia for failed or incompetent managers.

A number of developments in the past 10 or 15 years, however, seem finally to have drilled into the consciousness of American management its failure to achieve excellence or even competence in responding to the needs of American workers. One critical incident that shook the self-esteem of American managers was the failure of Motorola's Chicago television production facility and its subsequent success under Japanese managers. In 1974 Motorola finally admitted that it could not successfully operate its TV manufacturing plant in the face of international competition. Defects in the final product were high, and labor productivity was low and getting worse. The facility was sold to Matsushita, a Japanese television manufacturer, and Japanese managers came to Chicago to reopen the plant. Using the same American workers and the same physical facilities to produce the same product, quality and productivity began to rise, until by 1976 the plant was producing at quality levels that rivaled the best factories in Japan.

Long after it was obvious that authoritarian management practices had lost their effectiveness, they continued in use by stubborn American managers. But the Vietnam experience encouraged younger workers to resist authoritarianism in the workplace. GM had major problems at its Tarrytown, New York, plants and at its new plant in Lordstown, Ohio. Tarrytown was old and had long been a center for militant unionism, but the Lordstown facility was new and in Ohio, considered the center of small-town, middle-American values.

Leading American managers became more concerned when they observed the directions taken by union movements in Germany, England, France, and Sweden. Manufacturers realized that they might lose control of the workplace unless changes were made. These concerns were reemphasized by the recession of 1980–83 and the consequent downturn in manufacturing. A "deathbed" conversion occurred to Japanese management style by many American manufacturing managers. Business consultants had more work than they could handle in catering to this latest management fad. Beneath the hoopla, however, major and permanent changes are under way, and they are explored in this chapter.

It would not be correct to think of participative management concepts as entirely new, because many of the best ideas in participative management have existed a long time but were ignored until recently. It would be equally incorrect to believe these effective tools exclusively Japanese in origin, because many elements in the Japanese style of management, such as quality circles and just-in-time inventory control have their roots in past American practice.

The chapter begins with a discussion of several major theories of management style: Theories X, Y, and Z. Theory X is the name given, after the fact, to "scientific management" as proposed around the turn of the century by Frederick W. Taylor. It is a straightforward, hierarchical style and is still the most widely used method in American industry. Theory Y is a more participative approach advocated in the 1950s by Douglas McGregor of the Sloan School of Management at MIT and based on research by many individuals over the preceding 20 or more years. It is more in tune with modern psychological thinking than Theory X and has received widespread academic support. Theory Z is the name given recently by W. Ouchi to his more diffuse and vaporous presentation of what he takes to be the Japanese style of management. Ouchi sees Theory Z as a logical progression from X and Y [2].

Theories X and Y are well known and possibly of some importance in lending historical perspective, but from a practical point of view these static portrayals are limited, because they convey the false idea that management style is a matter of a fixed behavior pattern or even that it is a matter of the manager's own personality matrix. A more useful point of view, in my opinion, would acknowledge the self-evident fact that management style should be situationally based and dependent on the social maturity of the worker, as well as that of the manager.

A manager should manage differently with different individuals in different situations and even differently with the same individual under different circumstances if optimum results are to be obtained. The Hersey-Blanchard (HB) four-mode theory captures these ideas and is demonstrably more flexible and powerful than the older and more limited static conceptions. Thus after Theories X, Y, and Z, we shall discuss the HB four-mode theory and two other elements of participative management, quality circles and the Scanlon plan.

10.2 THEORY X

Frederick Winslow Taylor could not have been an easy man to know or like. He was opinionated, domineering, deficient in modern interpersonal skills, brilliant, hardworking, impatient with subordinates and superiors alike, and stubborn. He was born in Philadelphia late in the nineteenth century of well-to-do parents and was expected by them to go to Harvard Law School. He resisted this idea and at age 18, after an illness possibly of psychosomatic origin, his father permitted him to become an apprentice at the Enterprise Hydraulic Works, a small machine shop in Philadelphia owned by family friends. Taylor was bright and determined. It quickly became clear to him that the machinists of the day had little or no overall understanding of their craft and were quite inefficient [3].

Four years later, his apprenticeship complete, Taylor took a position as a common laborer at the Midvale Iron Works of Philadelphia also owned by family friends. Six years later he was the chief engineer at Midvale. Extraordinarily rapid as was his rise, it is universally regarded as due his genius rather than favoritism. Although Taylor observed conventional production norms as a worker, as soon as he accepted management responsibility he made it clear that he expected a day's work for a day's pay from those reporting to him. Three years of strife followed as Taylor imposed his will on the recalcitrant work force. The counter arguments some of his co-workers used and threatened to use on young Taylor's head and limbs after work made an impression on his body, but not his will. He rose to foreman and then to shop general manager. He more than doubled productivity at Midvale, but at some psychic cost to himself and considerable aggravation to the workers; profitable aggravation for all to be sure.

From this controversy, which Taylor detested, but from which he deeply believed he could not flinch, he came to understand that in some measure, the adversarial relationship between workers and management was due to universal ignorance of what could be reasonably expected of the typical shop worker when assigned a specific task. To eliminate this source of controversy and promote the worker-management cooperation he strongly advocated, Taylor began to conduct experiments on optimum methods of machine tool operation and on the composition and tempering of tool steel itself. This scientific experimentation was unique for its time, and its value was quickly recognized. But Taylor's subsequent work on job scheduling in the shop was even more revolutionary and controversial.

Taylor soon came to understand that the scheduling of jobs through a machine shop held great promise for significant improvements in productivity, and he began developing the theory of such scheduling. Up to this time, the American machine shop had always been operated on a personal basis by foremen who had hardly any more knowledge than the workmen they supervised. Hours of work, procedures, methods of work, schedules, and everything else were informally set and seniority, rather than skill, determined who did what job. Although the work was performed in a factory, the system was really a continuation of the old guild system of master workers, journeymen, and apprentices. Indeed when Taylor started his career, articles of indenture were still often required of new employees.

Separation of work planning from execution became a core element of Taylor's philosophy. It cannot be denied that such separation did promote efficiency, but it was and continues to be controversial because it is viewed by some as a surrender of individual autonomy. Others feel it helps workers to achieve their best and maximize income. To illustrate this point, Taylor used his famous analogy of the surgeon in training learning proper operating techniques from master surgeons. Individuality is not desirable in the apprentice surgeon learning a standard operating procedure, any more than would be the ''autonomy'' of a symphony orchestra musician be rewarded who played Brahms when the leader is conducting Beethoven.

It isn't difficult to imagine the impact of Taylor as he attempted to change every element of shop operation. He was determined to dictate which job would go on which machine and in what order. He said who operated the machine, at what speed it would operate, and the exact procedures to be used. He performed quality inspection when the piece was finished and decided if it was to be accepted and paid for. Taylor was determined that raw material and scrap be controlled as well as workplace discipline. In effect, he treated skilled machinists as raw apprentices.

The results were spectacular. Productivity increased in some cases more than tenfold. Taylor was responsible for taking the machine shop out of the guild system of the Middle Ages and helped create the modern factory system. His work on tool steel and cutting speeds became internationally known through his publications in the *Transactions of the American Society of Mechanical Engineers*, of which society he was elected president. Soon after becoming general manager, Taylor left Midvale, and after several contentious years as an independent consultant he was persuaded to accept a management position at Bethlehem Steel. Taylor was straightforward with everyone at Bethlehem concerning the methods he intended to employ. He had a right to assume that the company's owners understood his very plain English and were behind him. But the reality was different.

Taylor was dogmatic not only with the workers under him but also with the owners of the company. He was the epitome of the engineering specialist. He worked hard and he knew what he was doing. Taylor was perfectly prepared to acknowledge this to all comers and he resisted absolutely any attempts at interference. Ayn Rand would have loved him. Top managers at Bethlehem grew to hate him almost as much as did most of the foremen and many of workers.

The problem was that he was making everyone rich. Taylor advocated piecework plus generous bonuses based on production rates he determined personally. These rates were set to permit workers to increase their best take-home pay significantly, provided they followed Taylor's directives. At the same time production increased at an even faster rate, thus enriching management. But everyone had to do exactly what Taylor dictated, and no one enjoyed that.

Eventually the acrimony grew so intense that Taylor left Bethlehem to set up as an independent consultant and technical writer. Because he was a true zealot, he relished the opportunity to carry the gospel of scientific management as he called it, to the world. Taylorism gained more publicity than would otherwise have been likely, by virtue of the railroad-rate litigation efforts of Louis Brandeis and Theodore Roosevelt. Brandeis used the theory of scientific management in his arguments before the Supreme Court, and Taylor was asked to testify before congressional committees. The U.S. Navy began to use Taylor's approach in naval yards before World War I, and other nations grew interested. Taylor was an effective advocate because he had independent means and was more interested in making converts than in amassing further wealth.

He recruited acolytes such as Gantt and Galbreth to the true faith, but, as many religious leaders, he lived to see a number of his followers split off as schismatics. Taylor was soon spending as much effort fighting heresy as he was spreading the true faith among nonbelievers. Taylor suffered blinding headaches all his life, had several nervous breakdowns, and died young. His life and management methods provide much to intrigue the behavioral psychologist [4].

In considering Taylor's theories, keep in mind the spirit of the age in which he worked. Don't expect a 1990s level of social consciousness in a worker of the 1890s. Yet some modern theorists project modern social standards and theories on the social structure of 1900 and use them to impale Taylor. At the same time, it must be admitted that Taylor sometimes seemed deliberately to use provocative language that taken on its face, and out of context, could be viewed as offensive [5].

Taylor seemed at times to defy his hearers to penetrate his surface verbiage to learn what he really meant. This tactic requires a sympathetic audience, which Taylor rarely had. Consider the story of Schmidt. When Taylor moved to Bethlehem, he found that each laborer was expected to bring his own shovel and work with that shovel independent of the material he was handling. Taylor also found workers loading pig iron into rail cars without detailed guidance.

Given Taylor's zeal, we can almost see him smiling and resolving to change things. At this point in the story, a surface reading tells us that Taylor recruited a dull brute immigrant, code named "Schmidt," who had difficulty with English and who approached the caricature of a loutish brute. Taylor consciously or unconsciously promoted this caricature by calling Schmidt stupid and comparing his temperament to that of an ox [6, p. 59]. The deeper reality, however, is that Taylor carefully observed the approximately 75 laborers available and sought to enlist voluntarily the best worker in the shop. Taylor knew he would need cooperation and he went seeking this. One reason Taylor recruited Schmidt was because, while many workers headed directly for the local bar after work, Schmidt hurried home to construct a house for his family. Rather than a dull serf, Schmidt was the prototypical get-ahead moonlighter.

Taylor himself gives us the dialogue he held with Schmidt in recruiting him [6, pp. 41 ff]. He offered Schmidt the opportunity to earn a 60 percent wage increase if he would help Taylor determine the proper work pace and the timing of rest periods. Schmidt quickly volunteered. Taylor in his time and motion studies was careful to set rest breaks and to find a comfortable, sustainable pace. He did not want to strain or burn out workers, although such is his image. In fact it was alleged that Schmidt died under the strain, and years later Taylor was required to collect and present evidence that Schmidt was alive, well, and working. There seems little doubt, however, that some of Taylor's followers did on occasion set piece rates that required superhuman effort to achieve. Taylor knew this perversion would promote strikes and labor unrest, and he warned against making this error in his classic [6].

Yes, Schmidt was a guinea pig, but of the same sort modern professors of psychology themselves use, for example, a volunteer to whom the whole experiment is explained in advance. Even the most advanced protocols for human experimentation ask no more. Taylor and Schmidt learned the proper speed at which to unload pig iron from rail cars and the proper timing of rest breaks. Taylor then went on to test and select the best shovels to use for handling foundry sand, coke, and ore. Taylor used these homely examples in his speeches and writings but was disappointed that some critics claimed this was all there was to scientific management.

It is interesting that the U.S. Army and Navy picked up on scientific management faster than private industry prior to World War I. One is tempted to argue that the sense of hierarchy and discipline embedded in Taylorism appeals to the military mind. This may be another modern mistake, however, because captains of industry at the turn of the century were generally just as domineering as the military. Industry's early neglect of Taylorism may have been based on the arrogant way in which it was presented, or it may simply have been bad management.

When industry did choose to try scientific management, it sometimes did so in a way that gave these techniques the bad name they have long suffered. Contrary to some of the time-and-motion technicians who followed him, Taylor strongly advocated sharing with workers the benefits of increased productivity gained by following his methods. He set fair piece rates that greatly increased the earnings of skilled workers. But greedy shop managers soon began to resent these higher earnings and cut the piece rates. Workers learned that if they started to break the rate and increase production, the rate would be reduced. Instead of the partnership or implicit contract Taylor envisioned between management and workers, many shop managers retained the adversarial approach and used efficiency experts and time-and-motion people as spies. This perversion of Taylorism no doubt hastened the onset of the organized labor movement.

Taylor's scientific management consisted of the following seven elements:

1. Separation of work planning and scheduling from the execution or work effort itself

2. Careful measurement of the time required and the most efficient motions to be utilized, along with proper rest cycles to be taken by the worker so as to realize maximum effective results with minimum effort

3. Careful design and maintenance of workplace, tools, jigs, and fixtures and designation of their proper use

4. Sharing with the worker the results of improved productivity through fair and fixed piece rates and high bonuses

5. A cooperative relationship between workers and management

6. Consideration of the worker as an individual

7. Careful selection, training, and reward of willing workers and demotion or elimination of unwilling, unskilled, or unsuitable workers

The modern reader may ask how a cooperative relationship and consideration of the worker as an individual can be maintained at the same time as total control of the workplace is asserted and the right to fire unsatisfactory workers is retained. Perhaps the answer lies in the way the human relationships are built and maintained. Taylor considered cooperation desirable, but it is clear that he started with a work-centered view rather than a worker-centered view. He expected the worker to conform to the needs of the job, not the reverse. As unpalatable as it may be, termination of employment as an ultimate sanction is part of all management methods.

10.3 THEORY Y

By the 1920s the factory system was deeply embedded in America, and the concept of the worker as an independent entrepreneur or guild member had disappeared. The American Telephone and Telegraph Company was big and rapidly growing bigger. AT&T had a large Western Electric plant at Hawthorne, Illinois, outside Chicago, which produced telephone parts and electrical relays for central office switchboards and the like. Hawthorne production engineers and staff industrial psychologists began to look into assembly methods used by the relay workers, all women, just as Taylorism dictated. Perhaps these male professionals were more polite to the female workers, or perhaps they were generally more skilled in interpersonal relations than their counterparts in machine shops and foundries. In any case, the workers did not see these professionals as enemies.

As the investigators began to study the ideal bench height, seating arrangements, and parts placement for best assembly, they listened to the advice of the relay worker teams. Sometimes the relay workers discussed changes on their own and suggested them on the engineers' next visit. Meanwhile, production increased in the teams being studied. Control teams in other relay rooms were left alone, and their production did not increase. In 1924 studies were begun on the effect of lighting intensity on production, and a positive correlation was found between productivity and light intensity. Thus far this was completely orthodox scientific management.

The next portion of the lighting experiment, however, contradicted Taylor, and the way the contradiction was handled demonstrated scientific excellence. The contradictory results were not ignored or buried. The investigators noted that as lighting intensity was reduced, productivity continued to climb! In fact, even when intensity was reduced to a level below that of the control group, production in the experimental group continued to climb. Taylorism's intense focus on the workers' tasks and absence of attention to workers' feelings doesn't explain this phenomenon.

Early in the course of these ten years of experiments, Western Electric personnel mentioned their results to Elton Mayo, a professor at the Harvard Business School, who encouraged them to continue their work. In fact, Mayo helped raise money to keep proper statistical data on the results and persuaded Western Electric management to let the experiments continue. Years later the illumination experiment data and experiments in coil winding and other processes at Hawthorne were published [7]. The response to the book *Management and the Worker*, by Roethlisberger and Dickson, was little short of astounding. It kept a whole generation of sociologists and industrial psychologists busy at their typewriters for decades.

Mayo felt that the Hawthorne experiments indicated that Taylorism was deficient because it treated workers as cogs in the machine and denied their humanity. Based on this interpretation, he advocated a style of management that concentrated on human relations [8]. Opponents of the Mayo human relations school denied Mayo's interpretations and extrapolations on the Hawthorne data. Indeed, the original Roethlisberger-Dickson data are ambiguous and not easy to grasp. The authors themselves admit confusion over the results. Landsberger's later monograph on Hawthorne and its aftermath is shorter and more accessible [9]. Landsberger points out that both the Mayo human relations school and its opponents have misinterpreted Roethlisberger and Dickson. Nevertheless, the Hawthorne study, confused, ambiguous, eclectic, and flawed in methodology though it may be, formed the base for much of Theory Y.

Douglas McGregor became interested in the Hawthorne results and was an immensely effective advocate given his faculty position and energetic writing efforts. He and his colleagues continued studies on effective personnel management for many years, and in his famous book, *The Human Side of Enterprise* [10], McGregor differentiated between Taylorism, which he called Theory X, and the more participative style demonstrated by the Hawthorne experiment, Theory Y. McGregor's book is small, easy to read, informal, and considered a classic, but it has its flaws. There is no index, but the references contain not a single mention of Taylor. This omission is striking in a text that's purpose is to propose an advance over the traditional Theory X style of Taylorism.

More important, however, is McGregor's mistaken emphasis on the bimodal contrast between two static concepts, the highly task-oriented Theory X and the highly interpersonal Theory Y. While McGregor recognizes the transient nature of management style (see his Fig. 1, p. 25), and predicts that education and increasing social maturity should bring management and labor into a natural partnership (even Taylor accepted this in the abstract), the main thrust of McGregor is to emphasize two static and opposing managerial world views, Theory X and Theory Y.

10.4 TOWARD THEORY Z?

In the 1960s Abraham Maslow, a distinguished American psychologist, became interested in industrial management and accepted an invitation from Andrew F. Kay, the president of a small California electronics firm, Nonlinear Systems, Inc. (NLS), to reorganize its management style. Maslow helped NLS adopt a thoroughgoing Theory Y posture and indeed pushed further than McGregor suggested [11, 12]. McGregor argued that Theory Y worked best with highly educated workers committed to the goals of the organization, [10, p. 55, pp. 62–74], for example, managers themselves and research workers in industrial laboratories. Maslow, on the other hand, encouraged NLS to install an advanced form of Theory Y on the factory floor with ordinary blue-collar workers. Kay fostered a familylike atmosphere at NLS. Production workers were involved in sensitivity and training group sessions. They were consulted on production decisions, and Kay clearly enjoyed the widespread publicity of his experiments in the workplace [13]. *Reader's Digest* ran an article, and McGregor took notice in a later book.

Initially it worked quite well. However, in the post-Vietnam recession experienced by the California aircraft and electronics industry, NLS began to lose money. In 1970 Kay decided it was necessary to revert to hard-nosed Theory X management style. A few industrial psychologists, not including Maslow himself, had made extravagant claims about the NLS experiment, and when it failed, Kay's management consultants turned on Kay and accused him of never really being a believer [14]. Maslow, however, had been cautious in his prognosis and indeed had predicted problems should a recession occur [11].

NLS is discussed here for several reasons. First, there is a general lesson for managers about consultants. Consultants, no matter how brilliant, do not have profit-and-loss responsibility in the concern. The manager, not consultants, must manage. In hindsight, Kay doubtless allowed theorists to persuade him to take NLS too far too fast and allowed theorists too free a rein. He recognized this and was quoted by *Business Week* as saying, "I have lost sight of the purpose of business, which is not to develop new theories of management" [14].

The next chapter in the NLS story negates Kay's carping personal critics. NLS weathered the problems of the early 1970s by strict attention to operational details and a tight task focus. Then in 1979 Kay redirected NLS to the personal computer market, renaming his company Kaypro after the portable computer that became its flagship product. Sales dropped from $6 million in 1965 to $3.5 million in 1971. Kay's hard-nosed changes lowered his break-even point, and the company showed small profits as sales crept up to $4 million in the next few years. But the situation turned sharply for the better in the middle of 1982 with the introduction of the Kaypro II. Sales in 1982 were $5.2 million, and they zoomed to $75.2 million in 1983. Kay feels he was able to take advantage of this sales explosion because of his high interpersonal management approach. Note the combination: high task orientation *and* high interpersonal orientation. This unusual combination will be discussed in a moment. Before doing so, however, we will attempt to assess a theme in management style that swelled to become a dominant chord in the late 1970s and early 1980s: the so-called Japanese style of management.

As an example of this trend, take Ouchi's book on Japanese-style management, which he calls Theory Z [2]. Ouchi describes approvingly a highly interpersonal, family-style, participative industrial environment. As did McGregor, Ouchi sees management style as a matter of the manager's static world view. He quotes McGregor's familiar X-Y dichotomy:

> McGregor felt that these [managerial] assumptions [concerning human nature] were primarily of two kinds, which he labeled "Theory X" and "Theory Y" assumptions. A Theory X manager assumes that people are fundamentally lazy, irresponsible, and need constantly to be watched. A Theory Y manager assumes that people are fundamentally hard-working, responsible, and need only to be supported and encouraged.

Ouchi is vague on the details of Theory Z, but he advocates stepping beyond the consultation and persuasion common in Theory Y into real partnership responsibility, involving the production worker in control of production quality and other tactical factory operations. Ouchi advocates a higher interpersonal approach and lower task orientation. Note an echo from the earlier 1960s days at NLS. However, Ouchi's approach contains no new elements. It is merely an intensification of the highly interpersonal relationship already present in traditional Theory Y.

Ouchi emphasizes the long-term employment relationship and the familylike or clan aspects of a Theory Z organization. An employee is adopted into the clan, or family, and orients toward it rather than toward the craft or the marketplace, growing and maturing within the organization for an entire career. Ouchi's warm, anecdotal, impressionistic approach to his topic is delightful, but his uncritical enthusiasm is disconcerting. Ouchi seems to be aware of some of the difficulties of establishing a high-intensity social fabric in the workplace isolated from the remainder of the social fabric current in western society, but he may underestimate their weight. However, Ouchi's theory may provide a clue as to why the high hopes for Theory Y held by theorists for almost a half century have so often been dashed in practice.

The social fabric of the Eastern world is more tightly knit than that of the West. Eastern culture is more primitive and more oriented to the clan and family. Widespread social obligations across the generations and throughout the extended family are the common mode. Thus the apparently more advanced Theory Y is actually more in tune with traditional values and current world view of the East. Theory Y is not necessarily doomed to fail in the West, but its introduction and application should be reassessed.

The philosopher Hegel proposed that logical argument, or dialectic, should proceed by first stating a *thesis,* then an opposing argument, or *antithesis,* and finally the two should be blended together, retaining the best of each, *synthesis* [15]. Suppose we try this with Theories X and Y. Theory X is highly task-oriented and low in human relations content. It focuses on economic efficiency and individual effort and reward. Theory Y is the antithesis. It takes a low task orientation and a high human interrelationship orientation. It focuses on human equity and emphasizes communal and cooperative elements of the work environment. Ouchi's Theory Z is not a Hegelian synthesis. It is merely a modernized presentation of Theory Y with what is popularly supposed to be a Japanese flavor. Nevertheless, a Hegelian synthesis of Theories X and Y does exist, although its creators do not recognize this perspective.

10.5 THE HERSEY-BLANCHARD FOUR-MODE THEORY

Hersey and Blanchard [16] have developed a theory of management style that breaks away from the static bimodality of Theories X and Y to produce a four-mode theory of management style based on the social maturity of the group to be managed. This dynamic four-mode theory seems to me to have a solid psychological base in that it focuses on the interpersonal context of the managerial situation. Furthermore, the theory can be used to provide a framework for analysis of the less contextual, and therefore more elementary, Theories X and Y. The Hersey-Blanchard (HB) theory accepts the dynamic nature of the interaction between worker and manager. In addition, workers can employ the HB theory to analyze supervisors and manage their mutual interaction. That is, the worker can manage the manager for the overall good of both and the organization. Most valuable of all to the manager is the predictive ability of the HB approach. Properly employed, it serves as a guide and a feedback tool for improving manager and worker effectiveness.

Starting with the mode appropriate for the lowest level of worker maturity, the four modes of Hersey and Blanchard are labeled "tell," "sell," "participate," and "delegate." The "tell" mode (T mode) of management is a strongly hierarchal, directive style directly comparable to the Theory X style advocated by Taylor. It is highly task oriented and low in interpersonal intensity. It is used in basic training of military troops and is common in heavy industry in the United States. It is considered necessary when workers are not mature and cannot be relied on to cooperate with management for the benefit of both. Psychologically, the T mode assumes that workers are children who do not know what is in their own long-term best interest.

Given the social conditions in the nation and the state of factories of America at the turn of the century, one can see the applicability of this method. Many workers were immigrants who did not know English and were unused to the discipline and organization necessary in the factory. They were farmers from Europe completely without industrial training. Even expert workers had often been trained in the guild system, which featured individual craft skills rather than the more organized approach needed in the modern factory.

The next, more advanced mode is called "Sell" (S mode). In this case the manager takes time to explain why a particular approach has been chosen. The mode is still hierarchic, task oriented, and not fully based on the informed consent of the worker. It is the style used in leading skilled military troops and in many modern blue-collar task groups. It is higher in interpersonal intensity but not opposed to Theory X and indeed is an advanced version of the method. One can imagine Taylor stopping to explain to a bright and willing worker why a task is being done in a particular way. The manager who practices the sell technique is moving toward seeking the informed consent of the worker. Psychologically, managers in this mode see workers as advanced adolescents who generally can be relied upon to understand what is good for them but are not yet ready for full adult responsibility and complete independence.

"Participate" (P mode) is the third mode of management and may be compared to Theory Y. The worker is considered mature and willing to give informed consent. Furthermore, the worker is viewed as having good ideas that can be incorporated in the management plan once the goals and constraints have been made clear. In fact, the P mode's goal is to elicit tactical suggestions from workers rather than impose directions from above. White-collar workers and professionals should be in this mode. As a matter of fact, many factory workers have long been ready for this mode as well, as the success of the Scanlon plan and quality circles demonstrates. Psychologically, the P mode worker is a young adult, fully responsible and ready to participate in the decision process as far as training and experience permit. The P mode is lower in directive task orientation than the T mode but retains the high interpersonal flavor of the S mode.

The "delegate" (D mode) is the most mature mode of management. Management assumes fully self-actualized workers, ready and able to organize themselves to handle their own responsibilities. The D mode is similar to, but not the same as, the concept psychologists call the "leaderless group" mode of management. In the D mode, the manager and the work group have arrived at a common understanding of the organization's general goals, and the manager leaves the team to handle tactical and organizational details as well as daily operations to accomplish the goals. Management acts as a resource and a consultant rather than a task leader. In the D mode, the worker is seen as a motivated, mature adult with whom the manager transacts business. Table 10.1 shows the interrelations among characteristics of each mode.

TABLE 10.1 CHARACTERISTICS OF THE HERSEY-BLANCHARD FOUR-MODE THEORY

Mode Name	Interpersonal Relationship	Task Orientation	Worker Maturity Level Assumed
Tell (T mode)	Low	High	Lowest
Sell (S mode)	High	High	Moderate
Participate (P mode)	High	Low	Higher
Delegate (D mode)	Low	Low	Highest

Figure 10.1 shows the HB matrix and the trajectory of management style based on the maturity level of the worker. Each quadrant is uniquely defined by a combination of intensity of interpersonal relationship between worker and manager and intensity of task orientation. The bell-shaped curve is the trajectory along which the mature manager adjusts management style, in response to, and for the purpose of, increased socioemotional maturity level of the workers. This appreciation of environmental context makes the HB approach unique [16].

Almost all people seem automatically to resist change and managers can expect such resistance in their efforts to move forward along the HB trajectory. However, the same resistance may be useful through peer pressure in resisting the backsliding of a team member. Not all workers in a group are uniform in their maturity level, and not all react similarly to managerial stimuli. The manager must expect that a year or more may be required to move a group from one quadrant to the next. Some individuals and groups may never move all the way along the trajectory to the D mode.

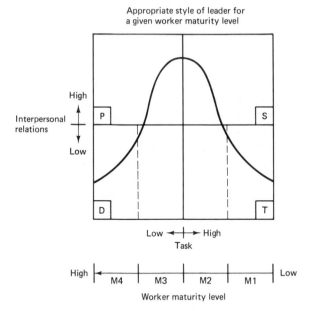

Appropriate style of leader for
a given worker maturity level

Figure 10.1   Hersey-Blanchard matrix.

Suppose that a group is operating well in the T mode. How do you move into the S mode? Start by supervising tasks a little less closely, reveal more of the overall plan, and allow a little freedom. Do not increase the intensity of interpersonal relations before this delegation is successful. From a point on the trajectory in the middle of the T mode, temporarily move off the trajectory to the left. Do not make a large change but rather one that is small but noticeable. When this change is successful, move up to the trajectory by increasing your interpersonal relationship. Become more interested and supportive of your worker's state of mind. Be a little less formal and more friendly, but be cautious. According to Hersey and Blanchard, workers sometimes come to view a manager who reverses these two steps as a soft touch.

Now suppose the group is operating well in the P mode and additional delegation is successful. Hersey and Blanchard suggest that the manager now withdraw somewhat by reducing his or her relational efforts. Would this be viewed as punishment, as it would in the S mode? Not if the workers are truly mature. They are happy in the P mode, but their self-actualization will be increased by leaving them more on their own.

Of course, any theory is an abstraction, and this theory is no exception, but it does provide a powerful illuminating effect. The manager attempts to move people along a maturity curve as rapidly as possible because mature transactions among adults are superior to other games people play. Naturally, not all individuals arrive simultaneously at the same maturity level. Furthermore, the manager must be mature to work with various groups at differing levels of maturity. A manager has to be alert to temporary slips in maturity. For example, unusual pressure on the job or family problems often correspond to a downward migration in maturity level. Because the Hersey-Blanchard theory focuses on the social maturity of the workers managed, it provides an indicator to the manager of the appropriate approach to a management situation and a check on the proper application through the response of the group. This guidance and feedback effect is not present in Theories X, Y, and Z.

Turn back now to the NLS case for further analysis using the HB four-mode theory point of view. From this viewpoint, the transition of NLS from Theory X to Theory Y was excessively abrupt and asking for trouble. A manager may need to dwell in the S mode for several years. At NLS opposition to the transition existed from the beginning but was overridden. Note from the description by Wolff [13] that now Kaypro has moved back from the T mode to the S mode, and Kay hopes the move will be permanent.

The HB four-mode theory is consistent with the quality circle concept and the Scanlon plan, discussed next.

10.6 QUALITY CIRCLES

The quality circle concept is only one element in a participative management environment. It has attracted the most attention because it is something that plant visitors and top management can see and thus is tangible evidence that something different is going on. But for the approach to be effective, the whole work environment must be supportive. A manager cannot install quality circles (QCs) in an otherwise strongly authoritarian Theory X environment. Yet that is just what some American managers have attempted.

Expect failure if management attempts to move precipitously from the T mode to the P mode (witness NLS), but precipitous moves were initiated in a number of QC installations that later proved abortive. The QC manager of a *Forbes* 500 company reported to me on the origin of QCs in his moderately authoritarian corporation. The CEO had decreed initiation of a QC program on a Monday, after seeing a segment on QCs Sunday night on *60 Minutes* on TV. Quality circles are small *voluntary* groups of workers, usually from the same plant section, that come together weekly during the workday to discuss the working environment and plan steps to improve that environment, the quality of the work product, and their own productivity. The circles must have management support to be effective. Trained facilitators must be present to organize the circles initially and to work continuously with the circles, their leaders, and management if the circles are to be effective.

The quality circle movement expanded rapidly in the United States. Fewer than ten companies were using them in 1975, but over 2500 circles were reported to be functioning in 1981. The reason for this enthusiasm is plain to see. Quality circles are a central part of the Japanese management system, and quality circles were reported to show an ROI of over 80 to 1 in Japan. More than 1 million circles are said to exist in Japan, and over $20 billion in cost savings are reportedly achieved there each year [17]. No doubt these estimates are based more on informal guesses by enthusiastic advocates than on careful analysis, but even so, the numbers are too large to ignore. Nevertheless, it would be surprising if more than a small fraction of the 2500 QCs reported active in the United States in 1981 were still operating five or six years later.

Here are some of the characteristics displayed by successful QC programs.

Management Support

The installation of participative management techniques and quality circles must be supported by management. Management must take the initiative in demonstrating tangible support to workers.

It is not at all uncommon for top management of a firm to give apparent support to the quality circle concept because it wants to get something for nothing. Management expects to make no fundamental changes in the way it operates, although it expects the workers to do so. If quality circles are started in such an unpromising environment, middle-level managers will feel the heat first and will help to precipitate failure.

I am familiar with a production plant of a moderate size U.S. corporation, in a relatively high-technology business sector that installed a QC program. The plant was fortunate in its (essentially accidental) choice of the supervising facilitator. The facilitator had never indicated any previous interest in QCs, but after only a few weeks' training was thrust into action. After a number of startup problems, the circles began to click, and one suggestion saved the company over $1 million in its first year. A year later, the successful facilitator was transferred to a new job, and the circles were dying. Do not underestimate the roadblocks that X-style middle managers can impose.

Voluntary Participation

Membership in a quality circle must be voluntary on the part of the worker, and at least 20 to 50 percent of the work force must participate for a significant impact to be realized. If only a small fraction of the workers are willing to try the QC concept, suspect either a badly alienated work force accustomed to operating deep in the T mode, inadequate preliminary discussions, or both.

Leader Training

Experience indicates that a supervisor or first-level manager can be a QC leader, but the individual chosen by the circle as leader must have special training. It is very difficult for first-level managers in a thoroughgoing Theory X organization to adopt the new techniques, but it can be done given sufficient time and encouragement. American auto companies such as GM are a case in point.

Focus

The QC group must stay focused on the agreed goals. It should not degenerate into a social club or a gripe session. Furthermore, the focus must be carefully chosen within the scope of work of the members of the circle. The circle was not formed to cast blame on other work groups or on management.

Quality of Work Life

Initially, at least, the focus should not be on increased productivity but on some aspect of improving the quality of the product, the process, the workplace, or customer services. These restrictions are admittedly artificial and can be waived as the circle gains experience. Management is not threatened by explicit consideration of productivity early on, but union officials often are.

Formal-Informal Agenda

Informally, the quality circle builds fellowship, encourages positive spirit, and develops mutual trust, but it should not do so at the expense of its formal mission.

Forbidden Topics

Some topics, such as piece rates and wages, may be forbidden by the collective bargaining unit.

Advanced QC

As the workers exhibit interest, they should be encouraged to learn how to self-inspect their work product and begin to apply statistical quality control methods.

Circles should meet weekly for approximately one hour on company time in a quiet room with chalkboard or flip chart. A professional, full-time facilitator should help organize circles initially and meet with them regularly. At first the facilitator leads the group discussion but must quickly get out of the leadership role. Leadership should come from within the group, possibly on a rotating basis. The facilitator provides training for the leader privately or in groups of leaders. To abandon the QC leader prematurely or omit preliminary and ongoing training is to ensure failure.

Leaders must learn to encourage discussion, be supportive and not argumentative or negative, and encourage participation by quiet members of the circle. Full discussion should precede choice of group focus, and the facilitator and leaders must be careful to pick specific, modest tasks initially to assure success with the first few tasks.

Circles should be encouraged to present their own reports to management, and at least initially the facilitator should arrange in advance that the recommendation is assured of management acceptance. Some circle programs have failed at this point when middle-level managers reject a good suggestion because it appears to invade their prerogatives. Even more common is a good report apparently accepted and then buried by management. The facilitator must be alert for such conscious or unconscious management evasions of responsibility. The first few good reports should be presented by the originating circle to other circles to encourage and instruct.

I still ache when I think of how management treated one quality circle when it delivered its first report. The QC program was new, and other circles were watching to see how this first proposal was treated. The circle was made up of production workers, and many bought new suits for the presentation. The concept was a good one, and they spent their own time preparing the charts and rehearsing. Some engineering managers were late for the meeting, and one or two didn't come at all. They were cold and unhelpful during the presentation and unresponsive afterward. The whole QC program died at this plant soon afterward.

Inevitably, there will be obstructionists and negative thinkers in some QC groups, and such workers present the facilitator with a difficult problem. Individuals should not be banished from a circle unless intensive coaching by the facilitator has failed to correct the problem and then only as a last resort.

Although productivity may not be the explicit focus of a quality circle, it is certainly management's indirect goal. Labor leaders, however, recognize other values such as a more informed, contented, and reasonable work force and an overall improved quality of work life. Management must recognize that it is on trial in the QC approach, and if it exhibits negative or coercive attitudes, the quality circle process will fail.

Union leadership is also on trial when the quality circle approach is tried in the environment of collective bargaining. The collective bargaining environment is legalistic and adversarial in essence because this is the basic thrust of the body of labor law and the interpretations of the NLRB developed over the past fifty years. Union leaders view their job as representing the needs and wants of the workers to management. Such leaders might easily interpret the quality circle movement as a management effort to finesse the union and relate directly with workers. No doubt this view gains support if management is clumsy and accidentally or deliberately leaves the union out of the planning loop.

The last few paragraphs seem to be rather pessimistic. They mention failure of quality circles through worker failure, management failure, and union leadership failure. Thus it seems useful to ask the following question.

10.7 DO QUALITY CIRCLES WORK?

By 1982 the first rush of uncritical enthusiasm for quality circles in the American manufacturing environment had cooled, and some objective findings had emerged. Over 1500 U.S. companies had established QC programs by this time and a number of them had already failed [18]. A study of QC effectiveness by Goodfellow indicated that 21 of the 29 companies he surveyed rated their quality circles as financial failures. They had produced measurable improvements that were less than their cost [17]. This early indicator prompted Imberman to make a wider two-year study of circles at 41 companies. Of these, 28, or approximately 70 percent, rated their QC programs as failures in terms of positive financial payback [18]. If three-quarters of all QC programs have failed economically in the United States, clearly major problems exist in the way they have been implemented.

R. E. Cole argues that rapid adoption of the quality circle concept by U.S. managers has many elements of a fashionable fad [19]. Just as every up-to-date manager had to have a computer in the 1960s and CAD-CAM in the 1970s, even without objective justification, so in the early 1980s everyone was for quality circles. Cole further points out that adoption of the QC concept appeared easy because it did not threaten supervisors' authority and because an overwhelming proportion of early adopting managers believed QCs could be superimposed on the existing management hierarchy without other changes in style or procedures. In fact, Cole's survey of 207 early adopters of quality circles in the United States showed that 94 percent so believed.

Despite the failure rate, some QC programs have been successful in hard ROI terms and have improved the general quality of working life. Some successful QC programs in the United States report ROI factors as low as 3, but others have ROIs as high as 100, with the average around 10, according to the Ingles' report [17]. Thus the lesson to be learned is not to avoid QCs but rather to learn to make them work. Imberman found four common shortcomings in the 28 failures he studied [18].

Poor Management Persuasion

Imberman credits one common failure mode to a casual, quick-and-dirty familiarization program followed by a cut-and-dried, routine application of formalities. Employees are quick to detect this lack of internal commitment to change on the part of management. Labor does not merely turn away when a casual attitude is observed but also suspects hostile management motives. William Roehl, assistant director of organizing for the AFL-CIO, was quoted as saying, "A number of well-meaning people believe that Quality Circles can lead to improvements in the workplace. But what they don't know is that they can also be part of a company's union busting strategy" [18].

Poor Supervisor Training

Workers find it difficult to talk with typical Theory X organization supervisors. Theory X managers don't listen well, and they are impatient learning interpersonal skills. Yet these are precisely the skills it takes to lead a quality circle.

Cold Management Style

Supervisors in Theory X organizations learn the attitudes they adopt toward workers by watching how senior managers respond to them. Imberman cites worker complaints about dirty restrooms, muddy parking lots, management refusal to listen, and unfair, arbitrary decision making as worker turnoffs.

Poor Employee Morale

The employment concerns just mentioned contribute to poor morale and a deeply suspicious work environment. If management is not ready to work on this problem, installation of QCs is useless. Westinghouse is one of the large corporations that is reputed traditionally to have had poor worker morale, and when it attempted to install quality circles, in the early 1980s unions resisted. Hugh Taque, an official of the United Electrical, Radio, and Machine Workers Union, was quoted as saying, "We fought Westinghouse to prevent quality circle programs. We feel the whole program is a sham, because what the company is really looking for are speed-ups and elimination of jobs" [18].

Perhaps the place to start installation of quality circles is not the factory floor but rather with professional staffers and middle-management levels. It is difficult to see how a middle manager can get blue-collar employees to make an emotional commitment to the company when the manager's own motivation is missing. Yet many well-educated, potentially highly motivated individuals are sullen and depressed in the Theory X environment. While not disputing Imberman's argument that unions helped impede the introduction of quality circles at Westinghouse, I would also submit that many middle managers at the Circle W ranch in the early 1980s were also unconvinced of real change in upper-level management attitude toward them, and thus contributed to the disappointing results. Toward the end of the 1980s new leadership at Westinghouse Electric had moved to correct this situation. The work environment and productivity improved substantially.

Failure to understand that permanent change in management attitude may be necessary, failure to involve the union in initial planning, failure to train facilitators properly, and failure to train and support the circle leaders ensure the destruction of any quality circle program.

10.8 THE SCANLON PLAN

Section 10.7 might seem discouraging with respect to the potential for success of participative management techniques such as quality circles in the collective bargaining environment. To correct this misapprehension, therefore, one of the earliest and most far-reaching participative management plans in the United States, developed by an official at the United Steel Workers of America will now be discussed. The Scanlon plan is far more ambitious and far reaching than the quality circle concept and preceded it in conception by almost 40 years.

During the depression of the 1930s, Joe Scanlon, president of a Steelworkers Union local, led a successful effort to turn around his mill, which was in financial difficulty. He was encouraged in this effort by Steelworkers' national officials. Success led Scanlon to join the Union's national office and to apply his approach to other steel mills in financial difficulty. Later the Scanlon plan was successful at several profitable companies outside the steel industry [10, 20].

The Scanlon plan involves employee "gain sharing," traditionally anathema to both management and organized labor in the United States, but it is much more than a simple profit-sharing plan. It involves a two-tiered committee system of employee participation in the management of the organization. The first tier of committees is similar to the quality circle approach in the way it functions, but there are several operational specifics that make a large difference. The second-tier committee is made up of the chairs of the first-tier committees plus all company line managers and needed staff executives. This group serves as a direct communication link between the circles and top management.

In addition to gain sharing, the second-tier committee and required top management participation make a real difference in the Scanlon approach. One similarity, however, is the heavy pressure placed on middle-level managers, who often attempt initially to continue to operate as though the company were in a Theory X mode.

Installation of the Scanlon plan often requires over six months of preparatory meetings and discussions with all levels of labor and management. Over 90 percent of all employees must voluntarily agree to participate before the plan is actually initiated. Notice the difference in this careful, deliberate approach and the common mode of slapdash quality circle installation in the United States.

The first major task is to agree on the basis on which additional profits brought about by the process will be shared. Workers are often surprised at the low rate of profit earned by their company when the books are first opened to them. Workers must be led to understand the concept of ROI and the need for the company to earn a return on its capital investment. While almost every Scanlon plan company has a slightly different plan for distributing the marginal gains, all have an essential feature in common: the gains bonus is distributed without exception to all workers on a proportional basis.

Care taken initially in developing the bonus plan will be repaid in worker confidence and support. It is vital that the basis for bonus computation remain fixed for a long period—years, in fact—because frequent changes will inevitably engender deep suspicion and probably precipitate failure of the program. Furthermore, it is dangerous to begin the plan in bad financial times, although that is how Scanlon himself started, because if the first few months pass with no bonus payments, the plan may lose worker support.

The first-level committees begin meeting, after training, exactly as in the quality circle approach, but there need be no committee agreement on what tasks jointly to consider. Rather, committee members are encouraged to make individual suggestions to the group on how to improve quality of the work product and productivity. Members write suggestions on a form provided, and the rules require the committee to consider each suggestion. The committee's own productivity is measured by the number of suggestions it generates. If the committee agrees after discussion to support a member's suggestion, the suggestion is passed to the responsible line administrator for comment. If the committee does not support the suggestion, it dies.

If the administrative office responsible agrees to support the suggestion, it informs the originating committee of the timetable to be followed for adoption and the committee moves on to other matters. If management does not support the committee suggestion, it must come back to the committee with its counter arguments. If the committee agrees with the counter arguments, the suggestion dies. However, if the first-level committee still supports the idea, it goes to the second-level committee for review. The first-level committee meets weekly, and the second-level committee meets once a month. A trained facilitator or administrator for the Scanlon plan is essential not only for training and support of the first-level committees but also for keeping track of the suggestions and scheduling the work flow and agenda of the second-level committee.

The second-level committee is chaired by the general manager of the plant, and its membership consists of the first-level committee chairmen and the managers of all line and staff groups of the plant. The meeting opens with a presentation by a senior officer of the financial performance of the company for the month just past and year to date, along with discussion of the variances. A first-time visitor to such a meeting cannot help being surprised by the presence of production workers at such a high-level management presentation and impressed with their knowledgeable participation in the discussion of financial results. The visitor would be hard put to detect any difference in the level of presentation in this first portion of the meeting and a meeting at which division managers were reviewing progress with top corporate officials.

Following the financial report, the second-tier committee turns to a review of contested suggestions. Usually the first-level committee does additional homework on the contested suggestions, and the presentations are polished and businesslike. The first-tier committee is not expected to do its own financial analysis of the proposal; the analysis is prepared by an appropriate staff person. Still, it is startling to watch a janitor or machine operator present a financial justification based on ROI or other standard financial criteria for a proposal. If the middle manager responsible for the initial veto can't meet the arguments, the plant manager will rule in favor of the worker's committee.

A visitor to one of these sessions, which are good-natured but intense, easily detects the pressure under which middle managers find themselves. A good proportion of the suggestions get to the second-tier by default. That is, middle management didn't respond either way in a timely fashion to the first-tier committee, so the suggestion moves upward. Then, under the gun and in public, the manager responsible must respond. One can understand how managers get into this bind because they must handle regular assignments as well as plant committee interrogatories. But this pressure is one reason for continuing worker commitment and the success of the plan. This required timely management response is missing from most QC programs.

In the final segment, possibly late in the afternoon, after it has worked straight through the lunch period, the committee hears the plant general manager review future plans for the division, including projected production levels and significant new orders or changes of any kind. The visitor might hear that several European distributors aren't doing well and will be replaced, that a large order the previous month wasn't as profitable as anticipated because costs were underestimated in the bidding process, or other issues. This is also the time excessive rework in one department might be mentioned, or a report might be made on the installation progress of a new computer to handle materials requirements planning and plant cost controls.

A visitor at one such Scanlon plan meeting was surprised to hear a worker, a sewing machine operator to be specific, argue to the general manager that the overtime pay total for the preceding month was excessive for the current level of production and to hear her point out that overtime could be reduced by better job scheduling in the shop. The general manager agreed and pointed out that a new computer had been purchased for just such a purpose. It is surprising, and reassuring to see a general manager responding to questions from the floor just as if he were at corporate headquarters being quizzed by a tough CEO.

The Scanlon plan isn't the final answer, and some companies have had Scanlon plans fail. Nevertheless, it is exciting and reassuring to those concerned about the capability of American managers and their ability to provide the kind of responsive leadership that American workers deserve.

EXERCISES

1. In Chapter 9, the Packard Electric Division of GM was described as having suffered a long-term deterioration in labor relations in its Warren, Ohio, plants. Is there evidence in the literature that any of the techniques mentioned in Chapter 10 might have helped?

2. Why did GE, Erie, go 100 miles south to Grove City, Pennsylvania, to construct its green-field diesel motor manufacturing facility?

3. W. C. Wood has suggested to the author that the sources of failure of the QC concept would make a fruitful study for organization theorists. In particular he postulates that Argyris's single-loop and double-loop learning concepts apply directly to this problem [21]. Discuss.

4. Why would a Scanlon plan worker object to excessive overtime payments?

5. Although Hersey and Blanchard do not mention it explicitly, workers can use their four-mode management style to manage bosses. Let's take an oversimplified example to illustrate the possibilities.

   A professor comes into class on the first day and hands out blank cardboard squares and marker pens. She says, "I'd like to ask you to make out these name tags and put them before you in class from now on so that we can get to know one another a little more quickly."

   The professor is in the _____ mode. Explain.

   To *confirm* the professor in that mode, the alert student will do the following:
   a. "Forget" the name tag the next time
   b. Say nothing but follow directions
   c. Smile, compliment the idea, and suggest adding first names to the tags
   d. Other

   Now the situation is the same but the professor comes into class and asks for suggestions on how to break the ice. She then offers the name tag idea and waits.

   Professor is in the _____ mode. Explain.

   Given that the professor is in the mode just determined, how would you attempt to take the professor to the next more mature management mode?
   a. Gently and in the kindest way possible explain why the name tag concept is a very tacky idea
   b. Smile and say the name tag is a good idea
   c. Do as you are told
   d. Suggest a classroom discussion to get additional ideas
   e. Suggest that the class take on the task of coming up with more good ideas and reporting at the next meeting

6. Suggest three specific ways in which the Scanlon plan differs from the quality circle approach.

7. The success of the New United Motors Manufacturing, Inc., NUMMI plant in Fremont, California, seems to confirm the Motorola, Chicago, TV manufacturing experience of the 1970s when it was purchased by Matsushita. In early 1982 GM was forced to close its Fremont plant because of high absenteeism, low product quality, and low productivity. In December 1984 a Toyota/GM joint venture reopened the old facility with many of the same workers. NUMMI productivity is about double the old figures, absenteeism has dropped to one quarter, the autos produced cost $1,000 less to produce than comparable GM models and they are GM's highest quality cars. What happened at NUMMI? Cite the literature and give a reasoned interpretation. See for example, Refs. [22, 23, 24, and 25].

REFERENCES

1. J. A. Young, "Quality: The Competitive Strategy," *Science,* November 4, 1983, vol. 222, no. 4622, p. 461.

2. W. Ouchi, *Theory Z: The Art of Japanese Management* (New York: Bantam, 1981), p. 69.

3. F. B. Copley, *Frederick W. Taylor: Father of Scientific Management,* 2 vols. (New York: Harper & Row, 1923).

4. S. Kakar, *Frederick Taylor: A Study in Personality and Innovation* (Cambridge, Mass.: MIT Press, 1970).

5. H. J. Leavitt, *Managerial Psychology,* 4th ed. (Chicago: University of Chicago Press, 1978).

6. F. W. Taylor, *Scientific Management* (New York: Harper & Row, 1911).

7. F. J. Roethlisberger and W. J. Dickson, *Management and the Worker* (Cambridge, Mass.: Harvard University Press, 1940).

8. E. Mayo, *Human Problems of Industrial Civilization* (New York: Macmillan, 1933).

9. H. A. Landsberger, *Hawthorne Revisited* (Ithaca, N.Y.: Cornell University Press, 1958).

10. D. McGregor, *The Human Side of Enterprise* (New York: McGraw-Hill, 1960).

11. A Maslow, *Eupsychian Management* (Homewood, Ill.: Irwin, 1965).

12. A. H. Kuriloff, *Reality in Management* (New York: McGraw-Hill, 1966).

13. M. F. Wolff, "Riding the Wave," *IEEE Spectrum,* December 1984, pp. 66–71.

14. "Where Being Nice to Workers Didn't Work," *Business Week*, January 20, 1973, pp. 99–100.

15. Georg W. F. Hegel, *Philosophy of the Right,* trans. T. M. Knox (Oxford: Clarendon, 1942).

16. P. Hersey and K. Blanchard, *Management of Organizational Behavior,* 4th ed. (Englewood Cliffs, N.J.: Prentice-Hall, Inc., 1982).

17. S. Ingle and N. Ingle, *Quality Circles in Service Industries* (Englewood Cliffs, N.J.: Prentice-Hall, Inc., 1983).

18. W. Imberman, "How to Make Quality Circle Programs Work," *Manufacturing Midwest,* May-June, 1982.

19. R. E. Cole, "Diffusion of Participatory Work Structures in Japan, Sweden, and the United States," in P. S. Goodman et al., *Changes in Organizations* (San Francisco: Jossey-Bass, 1982), chap. 5.

20. J. K. White, "The Scanlon Plan: Causes and Correlates of Success," *Academy of Management Journal*, 22, no. 2 (1979), 292–312.

21. C. Argyris, "How Learning and Reasoning Processes Affect Organizational Change," in P. S. Goodman et al., *Changes in Organizations* (San Francisco: Jossey-Bass, 1982), chap. 2.

22. M. Sepehri, "Car Manufacturing Joint Venture Tests Feasibility of Toyota Method in U.S.," *Industrial Engineering*, March 1986, pp. 34–41.

23. J. Schwartz, "Detroit's New Mentors in Managing Americans—the Japanese," *International Management*, September 1986, pp. 81–87.

24. D. Forbes, "The Lessons of NUMMI," *Business Month*, June 1987, pp. 34–37.

25. M. Parker and J. Slaughter, "Management by Stress," *Technology Review*, October 1988, pp. 37–44.

PART

IV

Small Business Entrepreneurship

The viewpoint of the preceding ten chapters has been consistently that of the larger firm, in which the accounting convention of "a going business" is reasonable. However, in a small, new, entrepreneurial endeavor, this assumption may be presumptuous. This will not require contradicting anything said previously, but rather certain additional assumptions concerning cash flow must be added.

The mental attitude of the entrepreneur is even more important. Whereas the employee of a larger concern can survive for a while without a display of personal initiative, an entrepreneur cannot. The entrepreneur must be a creator and a self-starter to survive.

Part IV provides an overview of the material in the previous parts, putting it all together. Each of the topics already discussed is of value in starting a new business, especially one that manufactures a product. The concepts are merely viewed from the entrepreneurial perspective.

CHAPTER

11

Small Business Entrepreneurship

11.1 INTRODUCTION

A small business differs from larger, well-established businesses in many important operational aspects and in one major financial accounting respect. The fiscal difference is illustrated by variance from one of the basic rules of generally accepted accounting principles, (GAAP). Financially speaking, it cannot be assumed that the small business is a going concern.

In a going concern, cash flow, accounts receivable, and accounts payable, along with inventory levels and other items, balance one another from quarter to quarter in a relatively smooth manner. Even if an established business is cyclic, with most sales concentrated in one or two quarters of the fiscal year, it is possible for it to establish a bank line of credit to carry over the lean season.

These things are not true in a small business. If sales fall off, even temporarily, the continued existence of the enterprise is threatened. While this fact may be obvious, it is not as obvious that a large new order can be not only a major opportunity but also a company-threatening crisis.

Even if a new business is growing and exhibits excellent profits in its monthly reports, it could be going under because of cash flow problems that never appear on the conventional financial statements or pro forma estimates we discussed so far. Cash flow is a more crucial financial control variable in the small business than in the larger, well-established concern. Cash flow and other financial elements especially critical to the small business manager form one of the topics of this chapter.

The second topic involves concepts of entrepreneurship and the fundamentals of starting a new business. Third, development of a business plan for the new enterprise is discussed. Writing a careful, objective business plan is important not only to impress possible funding sources but also to provide the needed focus to early planning efforts.

11.2 CASH FLOW

Conventional financial statements discussed in Chapter 5 are generally available in a firm and are certainly very useful. But more detailed figures are required for closer control, and different kinds of numbers may also be needed. Chief among these numbers is the cash flow of the firm. Cash flow must be differentiated from profit. Both of these concepts are precise, but they are different measures of the financial condition of the firm. Furthermore, the two are at best indirectly related. For example, a firm could show continuously rising sales and continuously rising profits from month to month and at the same time have an increasingly negative cash flow that could result in bankruptcy.

The cash flow condition of the enterprise is not mysterious and is easily computed, but because it is not one of the required financial reports, it could be neglected, to the firm's detriment. As Welch and White point out [1], a company roughly defines profit as revenues less expenses and cash flow as receipts less disbursements. Expenses and revenues are accounting concepts based on GAAP and paper records. They do not represent actual cash or cashlike transactions. Disbursements and receipts do represent real transactions. As Fig. 11.1 illustrates, a temporal difference exists between these terms.

- *Revenue* is "recognized," that is, entered on the books, when finished goods are delivered because then the customer is obligated to pay.
- A *receipt* occurs when the check arrives and is deposited and recorded at the bank.
- An *expense* occurs when material is used or consumed. Material cannot be expensed while it remains in inventory.
- A *disbursement* occurs when cash is withdrawn from the bank account of the business.

As a first approximation,

$$\text{Profit} = \text{revenues} - \text{expenses}$$

$$\text{Cash flow} = \text{receipts} - \text{disbursements}$$

Keep in mind that profit is an accountant's definition. It is a mathematical concept. It isn't money in the bank. Cash flow is real. It is money in the bank. Cash flow is time dependent. It matters when the check gets into the bank account.

Note from Fig. 11.1 that a lag of several months can exist between the time a check is written to pay for materials and the time the books record an expense. Significant lags can also occur between the time a revenue is recorded on the books and a check is received from a customer. These gaps contribute to the difference between paper profits and actual cash in the bank. If the loan officer has not been alerted to an increase in business that requires more inventory purchases, the business can easily run a negative

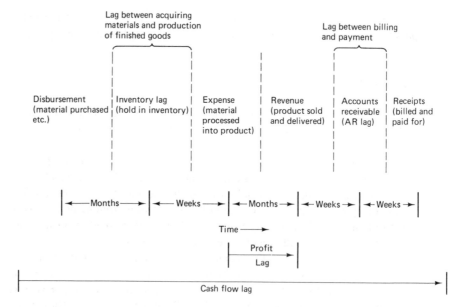

Figure 11.1 The temporal relationships among disbursements, expenses, revenues, and receipts.

cash balance that far exceeds its line of credit. Of course, if the company is a long-time customer with impeccable credit standing, the bank will listen to the explanation, but if not, the company could be in deep trouble just as business starts to boom.

As an example of the problem the lag in cash flow can produce, look at an abbreviated income statement and cash flow statement for a few months in the middle of the fiscal year at Ajax, Olympia. This new division of Ajax Metals is just starting to manufacture and sell sheet-metal watering troughs and similar items for the booming pleasure horse industry in the Northwest (see Table 11.1).

The pro formas in Table 11.1 look good, and the loan officer has approved the line of credit, to cover the negative end balance (EOM balance) for the first few months. These statements were approved in advance by the bank. The loan officer will be alert for variances, possibly on a quarterly basis, because the totals are small. Ajax, Olympia, is projecting a sales growth of 5 or 6 percent per month, which requires an increasing inventory investment. Sales receipts typically lag sales by about two months, causing this temporary negative condition. By Month 3, however, the end balance is projected to improve. If Ajax projected a few more months using the same assumptions, the balance would look even better.

But Table 11.2 shows the effect of higher sales growth on the ending balance. In Table 11.2 the rate of growth is 10 percent or better, and this requires more inventory. Because of the usual two-month lag in receipts, however, the balance is in bad shape, and if this rate of growth continues, the balance gets even worse two months later. Note the dramatically worse end balances for Months 2 and 3. This situation may or may not trigger a bank interrogatory, but it would certainly be embarrassing to explain.

TABLE 11.1   PRO FORMA STATEMENTS FOR AJAX, OLYMPIA

Pro Forma Income Statement

Item	Month 1	Month 2	Month 3
Sales	$176,000	$188,000	$200,000
COGS	104,800	111,400	118,000
G&A and marketing	50,740	55,940	59,340
Total expense	155,540	167,340	177,340
Taxable income	20,460	20,660	22,660
Tax at 30 percent	6,138	6,198	6,798
Net profit	$ 14,322	$ 14,462	$ 15,862

Pro Forma Cash Flow Statment

Item	Month 1	Month 2	Month 3
Sales Receipts	$160,000	$168,000	$176,000
Disbursements	148,200	152,200	166,300
Taxes	—	17,274	—
Total disbursements	148,200	169,474	166,300
Cash flow	11,800	(1,474)	9,700
Begin balance	(24,000)	(12,200)	(13,674)
End balance	(12,200)	(13,674)	(3,974)

Suppose Ajax fails to discover this point and doesn't bring it to the attention of the loan officer. It is possible that Ajax could be faced with an embarrassing ultimatum just as its business reaches critical mass. A company doesn't need that kind of trouble even if all eventually works out, and there is no guarantee that it will be all right. The two morals are (1) keep track of cash and (2) never surprise your loan officer.

The foregoing discussion assumes the standard form of accrual accounting. For very small firms, the simpler form of cash basis accounting will probably suffice. But a company should seek and follow the advice of a reliable accountant and tax expert because we cannot go into accounting formalities in this general introductory text.

Recall the usual two-month lag in accounts receivable, the time between billing and receipt of funds. The average age of the AR in a small firm is often longer than two months: 120 days is more typical. The entrepreneur needs to consider how to pay for food and lodging in the interim. A law requires that the federal government pay its invoices from small businesses within 45 working days or pay interest on the amount due

TABLE 11.2 IS AND CASH FLOW FOR AJAX, OLYMPIA, WITH HIGHER SALES GROWTH RATE

	Income Statement		
		Month	
Item	1	2	3
Sales	$176,000	$200,000	$220,000
COGS	104,800	118,000	129,000
G&A and marketing	50,740	55,940	59,340
Total expense	155,540	173,940	188,340
Taxable income	20,460	26,060	31,660
Tax at 30 percent	6,138	7,818	9,498
Net profit	$ 14,322	$ 18,242	$ 22,162

	Cash Flow Statement		
		Month	
Item	1	2	3
Sale receipts	160,000	168,000	176,000
Disbursements	148,200	162,800	177,300
Taxes	—	17,274	—
Total disbursements	148,200	180,074	177,300
Cash flow	11,800	(12,074)	(1,300)
Begin balance	(24,000)	(12,200)	(24,274)
End balance	(12,200)	(24,274)	(25,574)

from the date of the invoice. The government probably pays this penalty more often than it remits on time.

The new entrepreneur will enjoy Levin's text on the subject of management finance in the new, small business [2]. He approaches relevant issues in a lighthearted manner, but he is serious and his advice is excellent.

11.3 NECESSARY INGREDIENTS FOR ENTREPRENEURIAL SUCCESS

Five ingredients are necessary for entrepreneurial success. They are listed here and elaborated in the next few sections.

An Entrepreneurial Opportunity.   Entrepreneurial opportunity means more than having a bright idea. Furthermore, even the concept of a bright idea is not what most engineers and technical people think it is. It is necessary to take the viewpoint of the marketplace on the matter, not merely a brilliant technical concept.

Hard Work.   The key leaders of the new enterprise must be totally committed to the venture. They can't expect much time for their spouses and other family members, and don't have time for long weekends and vacations.

Adequate Funding.   Several guidelines are mentioned for obtaining adequate funding. The problem for most entrepreneurs is to obtain adequate funding and at the same time maintain control of their venture.

Careful Planning.   Both the business plan and the technical plan must be well done, and the leaders of the venture must watch for unintended variances.

Luck.   Who said life is fair?

11.4 THE ENTREPRENEURIAL OPPORTUNITY

An entrepreneurial opportunity involves more than just a nice concept for a product. The new product idea is vital, but it must be surrounded by overall product planning and aimed at a growing market segment. Planning should start with market demand, either overt or latent, and move back through product concept toward a set of specific product specifications. The product concept must be more than a patentable gadget. The total sales potential of the idea is critical. It must be a new advance in a rapidly growing market segment. Rapidly growing means more than 50 percent per year for the first few years. It is impossible to sustain a 50 percent growth rate for more than a few years, of course, but remember that at the beginning the base is very small. A good idea in a rapidly expanding market is better than an all-time great idea in a static field.

Contrary to the inclinations of the engineer-entrepreneur, a new idea in an old but growing field is probably better than a completely new product concept. Ideally, the competition should be a relatively old, expensive product that the entrepreneur can make better and cheaper through the use of new technology. This is better than a totally new product concept because a market demand does not exist for a new concept and must be developed.

Think about the position just presented. It seems contrary to common sense. Why isn't it better to move into an entirely new field of great promise in which there is no competition? The reason stems from the substantial time lag between conception of a wholly new product or service and its eventual success in the marketplace. This time gap can be years long and provides more than ample time in which to spend all available development funds and go bankrupt. Even large, well-established firms can be hurt by the cost of developing a new marketplace. Take the example of RCA. David Sarnoff was proud that RCA pioneered commercial TV, a great technical achievement. But Sarnoff was five or ten years too early, and RCA was seriously weakened by the financial strain of developing the market for home TV.

John Naisbitt's *Megatrends* [3] was on the best-seller list for months. As Naisbitt says, most if not all of his megatrends have been evident for years or even decades. Consider an early adopter who poured money, years ago, into a new product or service based on one of these trends when it first became apparent 10 or 15 years ago. The early adopter would not have accelerated the trend by much and would now be right but poor. This is not to say that trend forecasting is not important. It takes over a decade to reorient a large corporation, so an early lead is important in that environment. Individual entrepreneurs should not, however, get too far out in front of the parade.

Note that the bright idea must be demonstrably better and at the same time more economical for the end user than the best available alternative. Demonstrably means evident to customers as well as technically receptive specialists. If the product is merely better or cheaper, it is not good enough, because established competition can scale an existing product up or down to meet the challenge. Furthermore, it should really be more economical, not just made to look cheaper by reduced margins at various distributor levels or some such gimmick. An outstanding example of this double play of better and cheaper is the Japanese invasion of the British and U.S. motorcycle markets after World War II. Japanese manufacturers came into these markets with better, cheaper machines, and most of the sales they gained as the market exploded were to first-time buyers.

11.5 HARD WORK AND STRESS

The key personal attribute that distinguishes the successful entrepreneur from the rest of the world is psychic energy. Bright, well-trained people are turned out in gross lots by business schools, and young engineers with good technical ideas are also common. It takes more than these qualities to become a successful entrepreneur, and one additional element is an almost unlimited capacity for hard work. As one successful entrepreneur is reputed to have remarked to a group of business students, "You can become a successful entrepreneur by working just half days, six or seven days every week. And the beauty is, it doesn't matter which 12 hours of the day you choose."

Successful entrepreneurs are driven, willing to risk bankruptcy and careers on the strength of the conviction that they are right and the competition is wrong. Any venture will absorb all the cash it generates for the first year or so, so venture partners need some other source of personal income during this period, even if the venture is successful. This spreads the pressure to the family.

Most decisions an entrepreneur faces cannot be solved by analytic methods. Most tough decisions must be made on inadequate evidence, while the entrepreneur is tired, and under tight time pressures. Many decisions involve people. Often the decision comes down to whether or not to trust a person's word on little or no evidence. The entrepreneur must become a good judge of character.

Not only are all waking hours absorbed with business, but also so is the time and energy of family and close business associates. All family members should understand this fact at the outset.

Why does an entrepreneur do it? The answer is interesting. It isn't the obvious one, "To become rich." Rather, successful entrepreneurs seem to be driven by ego. They feel better, smarter, and quicker than the competition and obligated to prove it. They are extraordinarily competitive. To win and to do it on one's own is all important. An unusually large proportion of members of the Young Presidents Association are former Air Force fighter pilots, a superaggressive bunch of people.

The entrepreneurial life is a highly stressful existence, and the entrepreneur must cope with this fact. Evidence abounds in Silicon Valley and other venture centers in the form of broken marriages, chemical dependencies, and suicides, suggesting that many apparently self-confident entrepreneurs are in reality destroying themselves [4]. Consider an analogy to dieting and jogging. Maintaining weight and body condition by careful eating and exercise habits is a recipe for healthy living. An exciting, creative, enjoyable job is equally fulfilling. But some people, called anorexics, can't stop dieting even when it becomes life-threatening. Other individuals can't deny themselves the runner's high and run through pain and injury. Likewise, some entrepreneurs can't stop working and become workaholics. There is a considerable body of self-help literature on the subject, and it is worth taking seriously.

11.6 CAREFUL PLANNING

A bright idea in a high-growth business and a willingness to work hard are not sufficient. Entrepreneurs must plan carefully and watch every detail. If inventory gets out of control or attempts to collect accounts receivable fail the company will fail.

Suppose you and one or two other bright people have just quit good jobs at a large firm to try your own venture. It would be only natural if you were tend to work in the same style as before. But all the high costs invisible in a large organization can drag a small firm over the brink. An entrepreneur can't buy all the test gear available at Generous Electric or Ma Bell's Kitchen. You can't have all the backup inventory and a private office is no longer possible. The company principals have to live out on the factory floor when not calling on key customers or wooing bankers. MRP has to mean more now than a three-letter name for a computer program.

Entrepreneurs need to learn how to manage nonprofessional personnel as they add factory hands and how to build a network of distributors and keep them happy. The most important aspect of careful planning, however, is an area in which few engineers have had experience: financial planning. It is hard for the typical engineer, even the engineer interested in entrepreneurship, to accept that the company lives or dies by its financial results and not its technical excellence. The entrepreneur is playing the bankers' game and must play by their rules.

Entrepreneurs cannot closet themselves in the lab and let a hired bookkeeper run the front office. They must learn to understand costs and to keep costs under control.

They have accepted all general managerial responsibilities including long-range planning and are also responsible for major sectors of the business including finance, marketing, engineering, product design, manufacturing, distribution, and field servicing.

These special responsibilities can be delegated safely only to a venture partner who also owns part of the business and who will also lose everything if the company goes down the tube. An entrepreneur cannot delegate responsibility for strategic planning, however. Strategic planning is done poorly, if at all, in high technology. Silicon Valley is always in either a state of euphoria or a maniac low. No rational middle ground exists. In the winter of 1985 the Valley was on a new low. It was widely proclaimed and believed that the electronics industry was "in the worst depression ever faced." When translated, this statement turns out to mean that the industry was expected to grow at "only" 10 percent annually. Many industrialists would consider 10 percent growth heaven. But if a company hires personnel and acquires PP&E to cope with 50 to 100 percent annual growth, perhaps 10 percent does seem like the end of the world. On the other hand, respected business analysts had predicted this downturn almost to the month, more than a year in advance, but Silicon Valley managers weren't listening.

11.7 LUCK

Suppose you have served your apprenticeship and learned all you need to know to be a success. You have a great new product and several loyal partners. You have developed a great business plan and gotten venture capital. Early test models of your product work fine and dealers are willing to carry your line. What could go wrong?

Suppose the prime rate goes to 15 or 20 percent. What do you think is going to happen to that secured line of credit the bank offered to permit purchase of PP&E? The fine print probably says it is now unsecured. Suppose it is 1978 and the country goes into stagflation? Suppose it is 1981 and real business growth ceases and money stays tight, and no one could afford to buy any products, no matter how good they were. Or suppose one of the partners or favored suppliers turns out to be a crook? Suppose, suppose, suppose. The point of this litany of horrors is to say that an entrepreneur must keep some psychic distance from the venture even while giving it tremendous effort. Don't always be looking for the escape hatch, but don't let it destroy you if the venture fails.

11.8 SIX STEPS TO ENTREPRENEURIAL SUCCESS

In the mid-1980s the *IEEE Spectrum* organized a group discussion among successful Silicon Valley entrepreneurs to talk about how they did it [5]. From the discussion, the Spectrum editor gleaned the following six steps that lead to entrepreneurial success in high technology.

1. *Assemble the Team.* Venture capitalists aren't interested in talking to lone wolves with a good idea. They will help an entrepreneur find one missing specialist, for instance, a finance person, but most of the team must be on board and committed.

2. *Evaluate Yourself.* An entrepreneur needs breadth and steadiness as well as a capacity for leadershp and an enterprising bent. Be able to point to events in your life that demonstrate these capabilities. It isn't enough to imagine that you have them.

3. *Convince Investors.* There must be a clear, detailed conception for a good product and a good business plan. The business plan need not be a 100-page tome. Ten pages of facts are more than enough. But commitment is important. Don't expect someone to provide money for living expenses so you can quit your job and start a business with no pain and nothing but an idea.

4. *Slice the Ownership Pie.* Keeping 100 percent of the equity in the new firm and using debt funding sparingly takes time to build personal equity. Why not trade partial equity for time? Try to keep 30 to 50 percent, however, for yourself, your management team, and your employees.

5. *Start Working.* You can't figure out all the problems in advance, and the product's golden window of opportunity will close soon.

6. *Cope with Success.* Don't let the company crash after a painful startup and a few good years. Plan for continuity. Give employees stock options to increase loyalty and do forward product planning.

Incidentally, one of the entrepreneurs quoted in this article was Adam Osborne, founder of Osborne Computers, now bankrupt.

11.9 THE BUSINESS PLAN

Section 11.10 describes the concept of the new business "incubator." This is an apt name for an exciting concept. It is a place at which budding entrepreneurs are given help in bringing their ideas into profitable reality. A person who approaches the incubator director for a invitation to enter will be asked for a business plan. The director of one of the earliest and most successful incubators, created by George Low and George Ansell at RPI, said that few aspirants understand the importance of a business plan or how to write one.

Professor Fran Jabara, creator and director of the pioneering Center for Entrepreneurship at Wichita State University, and his able associate, Verne Harnish, founders of the Association of Collegiate Entrepreneurs, (ACE), in their courses on entrepreneurship, place primary emphasis on preparing a good business plan. The MIT Enterprise Forum, at regular meetings in Cambridge and spinoffs in New York, Washington, Houston, Chicago, and Amsterdam, spend the meeting time listening to and criticizing the business plans of new entrepreneurs.

All of these individuals and groups experienced in helping people start businesses and authors of most of the books on entrepreneurship concentrate on this same theme:

writing a business plan is the way to get started. The well-executed business plan is not long, or diffuse or unfocused. It is compact, terse, filled with numbers, and focused. It is the *only* salesman for the concept to venture capitalists and bankers, because they often won't give you time for a personal discussion without first reading the business plan.

Just any focus won't do. Rich and Gumpert, describing the operation of the MIT Enterprise Forum [6], give several possible foci, as follows:

- No focus: rambling, discursive
- Market focus: clients, customers, and competitors
- Investor focus: financial sources and owners
- Producer focus: product or service and how it works

Plans without focus don't get to first base. Because high-technology activity is associated with MIT, it isn't surprising to learn that many of the plans when first brought to the MIT Forum take a producer focus. Avoid the errors pointed out in our discussions of market pull versus technology push and don't make that same error.

Perhaps it isn't surprising that naive entrepreneurs, concerned with their ability to raise startup funds, often overemphasize the investor focus. The really successful plans, however, take a market focus.

The aspiring entrepreneur is well ahead on the process of preparing a business plan if the process of venture analysis is kept in mind. Venture analysis is the heart of the business plan. Adding three sections to a good venture analysis creates a complete business plan. Table 11.3 gives the section headings of a complete business plan.

TABLE 11.3 COMPARISON OF THE BRANDT OUTLINE OF A BUSINESS PLAN [7] AND THE VENTURE ANALYSIS OUTLINE IN CHAPTER 4

Section heading (Brandt)	Alternate title (Venture Analysis)
Overview	Executive summary
Concept	Description of venture
Objectives	Venture and corporate goals
Market analysis	Market potential
	Market opportunity
	Market segmentation
	Value-in-use (or competitive price)
	Market share
Production	Cost of sales
Marketing	(Market share, advertising plan, etc.)
Organization and people	
Funds flow	Pro formas
Ownership	

11.10 THE HIGH-TECHNOLOGY INCUBATOR CONCEPT

The new business incubator is one of those great ideas, beautiful in its simplicity, like the concept of the free public library, that return great multiples to society over and above their cost. The incubator concept harnesses the incredible energy of the free enterprise concept and turns its power to creating new jobs and a better social environment. In contrast, equally well intentioned government ''support'' programs often destroy in the long run the people they set out to help.

The incubator is a place where new businesses are created. Incubators support entrepreneurs before they are ready to find venture capital. The incubator provides a new entrepreneur with physical space and much more. To be admitted, the nascent entrepreneur submits a business plan to the incubator director. Often the applicant has a good technical idea but no idea of how to write a business plan and this is the first thing taught. More than half the applicants who profess an interest in starting a business never get around to completing the plan. Thus, in addition to many other useful functions, the business plan requirement serves a screening function.

In addition to low-cost space, the incubator often provides low-cost secretarial and bookkeeping services. Telephone reception, a photocopy machine, a coffeepot, and a conference room are also necessities. All venture leaders must agree to attend a weekly meeting chaired by the director. These communication meetings are used to share ideas, provide information, and give peer support. Occupancy is granted on a quarterly basis, and each venture leader must submit a quarterly report. These quarterly milepost meetings are more formal than the weekly breakfast or lunch sessions, and the incubator's board of directors should be present.

Many high-tech incubators are associated with nearby technical universities and can facilitate access to laboratories and libraries as well as to specialized advice from faculty and staff. Most startups have no credit rating and thus cannot purchase equipment or supplies. The ability of an incubator member to utilize the university purchasing service is a powerful advantage.

Another advantage is the showcase the incubator provides to potential venture capitalists. The incubator serves a validating and screening function for venture capitalists, and it is a great timesaver for them. To be able to visit and evaluate a dozen ventures in one day is a great help.

Obviously, the incubator concept is good for the budding capitalist with an interesting idea, but what's in it for the sponsor and the community? The community benefits directly by job creation. It also benefits by providing a home base for entrepreneurs. These individuals are high-energy leaders and create wealth. They are good people to have in a community. The sponsor can be a university interested in improving its surrounding intellectual and technical climate. The incubator is perhaps the only practical means of successful technology transfer from the university laboratory to the marketplace besides hiring the professors and students who create the ideas. Technology transfer is a personnel-embodied process. Following RPI's lead, hundreds of incubators have been created in almost every state in the nation, so the entrepreneur ready to launch should head for one.

EXERCISES

1. In the Brown article [8], it appears that Xerox may have made an error in rejecting John MacEachron's proposal to market LeaseAd software. Why did Xerox say no? Was Xerox correct?

2. It appears that MacEachron got a great deal from Xerox [8]. Enumerate the items Xerox gave MacEachron. Why was Xerox so generous? Are there benefits to Xerox from this arrangement?

3. Larry Crowley faced a standard new-business crisis [9]. How did he handle it? What are the pluses and minuses of his approach? How would you have handled it?

4. Aspiring entrepreneurs can learn from the experience of others, if they will. The post mortem of Osborne Computers in the literature is quite complete. Find the specific reasons for its collapse. Has Adam Osborne bounced back?

REFERENCES

1. J. H. Welch and J. F. White, *Administering the Closely Held Company* (Englewood Cliffs, N.J.; Prentice-Hall, Inc., 1980).

2. D. Levin, *Buy Low, Sell High, Collect Early, and Pay Late* (Englewood Cliffs, N.J.: Prentice-Hall, Inc., 1983).

3. J. Naisbitt, *Megatrends* (New York: Warner Books, 1984).

4. M. K. de Vries, "The Dark Side of Entrepreneurship," *Harvard Business Review,* November-December 1985, pp. 160–167.

5. T. S. Perry, "Six Steps to Becoming a Successful Entrepreneur," *IEEE Spectrum*, December 1982, pp. 46–50.

6. S. R. Rich and D. E. Gumbert, "How to Write a Winning Business Plan," *Harvard Business Review,* May-June 1985, pp. 156–164.

7. S. C. Brandt, *Entrepreneuring* (Reading, Mass.: Addison-Wesley, 1982).

8. P. B. Brown, "The Last Company Man," *Inc.*, July 1987, pp. 19–21.

9. L. Crowley, "Look, Ma, I'm a Businessman," *Inc.*, July 1987, pp. 96–97.

PART

V

Appendixes

APPENDIX A

Time Value of Money Calculations

A.1 THE FUNDAMENTAL RELATIONSHIP FOR COMPOUND INTEREST

The basic compound interest relation is the following.

$$F = P(1 + i)^n \tag{A.1}$$

where

F = future value after n periods ($)
P = present value ($)
i = interest rate (also called discount factor or discount rate, as a decimal, per interest period)
n = number of periods (not necessarily years)

Equation (A.1) is a simple relation, but others derived from it get somewhat involved. To reduce complexity, a simplified standard functional notation has been developed to represent compound interest factors. This functional notation is supplied for each relation. These functional relations cannot be manipulated, nor can they be solved or evaluated analytically. However, many tables are available for the standard factors, and these tables may be used to evaluate terms if care is given to round off error. The standard functional form for Equation (A.1) is

$$F = P(1 + i)^n = P(F/P, i\%, n) \tag{A.2}$$

Older texts, written before universal use of electronic spreadsheets, provide tables of the standard factors that often arise in compound interest calculations in order to

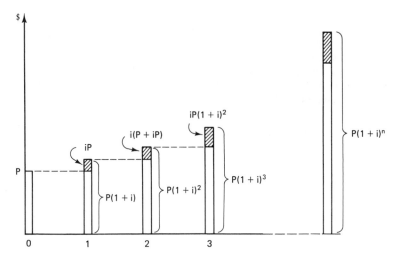

Figure A.1 Demonstration of a derivation of the fundamental compound interest relationship.

simplify hand calculation. One example is the value in parentheses in Eq. (A.1) raised to the nth power for which we have just given the functional notation in Eq. (A.2). This is the compound interest factor. Given the availability of the electronic spreadsheet, there is no reason to clutter the text with such specialized terms. For the hand-calculation approach, see any standard text, such as Ref. [1], [2], or [3].

It is not difficult to demonstrate how Eq. (A.1) is developed. Invest a present sum of money P at a rate of i percent per period for one period. By definition, at the end of the period iP dollars of interest is earned, and the total sum of money available for continued investment is the original principal P plus the accrued interest iP (see Fig. A.1).

$$P + iP = P(1 + i) \qquad \text{total sum at end of first period} \qquad (A.3)$$

At the end of the second period, $iP(1 + i)$ in interest has been earned, and the total sum of money available is

$$iP(1 + i) + P(1 + i) = (iP + P)(1 + i) =$$
$$P(1 + i)(1 + i) = P(1 + i)^2 \qquad \text{total at end of second period} \qquad (A.4)$$

In exactly the same manner, the total available at the end of the third period is computed:

$$P(1 + i)^3 \qquad \text{total at end of third period} \qquad (A.5)$$

By induction, Eq. (A.5) gives the future value $P(1 + i)^n$ after n periods.

Example A.1

Suppose that $500 is invested for three years at 4 percent interest. What is the future sum at the end of the investment period? By federal regulations, unless specifically told otherwise, a discount rate is assumed to be the annual nominal rate. From Eq. (A.1),

$$F = P(1 + i)^n = \$500(1 + 0.04)^3 = \$562.43 \qquad (A.6)$$

Using a three-place table of the compound interest factor to perform this calculation, a result of \$562.50 or an error of 7 cents is obtained. Although 7 cents is not significant by itself, it represents an error greater than one-tenth of 1 percent. Investing \$5 billion rather than \$500, the error would have been about \$700,000. Banks consider such a sum significant.

To make Eq. (A.1) more useful, rewrite it using very specific notation. Let

n' = Number of years

r = Nominal annual interest rate

r' = Effective annual interest rate

m = Number of interest periods per year

Then

$$i = r/m \qquad \text{and} \qquad n = mn'$$

Eq. (A.1) can thus be rewritten as

$$F = P(1 + i)^n = P(1 + r/m)^{mn'} \qquad (A.7)$$

Example A.2

Repeat Example A.1, but for quarterly compounding. Because the interest rate per period is the nominal annual rate divided by the number of periods per year and the number of periods is the number of periods per year times the number of years, we have the following:

$$i = r/m = 4/4 = 1 \qquad \text{and} \qquad n = mn' = 4 \times 3 = 12$$

Thus Eq. (A.6) becomes

$$F = \$500(1 + 0.01)^{12} = \$563.41$$

Quarterly compounding is worth about 0.2 percent more than annual compounding in this specific example. A three-place table lookup solution process would yield a result of \$563.50, or a 9-cent error.

A.2 CONTINUOUS COMPOUNDING

Because Example A.2 shows that quarterly compounding yields a higher effective rate of return, attention might be drawn to a bank offer of daily compounding or even continuous compounding. Let's look into this matter.

In Eq. (A.6), if m, the number of interest periods per year, becomes very large, approaching infinity, the term r/m approaches zero. This represents continuous compounding. To discover this limiting condition explicitly, write

$$F = P \lim_{m \to \infty} [(1 + r/m)^{mn'}] \qquad (A.8)$$

It is possible to reduce Eq. (A.8) to a standard form. If $x = r/m$, then mn' may be written as $(1/x)rn'$, and because as m becomes infinite, x approaches zero, Eq. (A.8) can be rewritten as

$$F = P[\lim_{x \to 0} (1 + x)^{1/x}]^{rn'} \qquad (A.9)$$

The value of the expression in brackets is well known:

$$\lim_{x \to 0} (1 + x)^{1/x} = 2.71828 \ldots = e \qquad (A.10)$$

where e is a transcendental number and the base of the natural logarithm. Thus Eq. (A.8) may be written as,

$$F = Pe^{rn'} \qquad (A.11)$$

Example A.3

Banks nowadays advertise continuous compounding, which has a nice ring to it. Instead of money sitting around in a bank vault and being compounded only once a year, investors can now visualize continuously compounded money in a state of constant activity even while one sleeps. A happy thought, but does it make much real difference? Suppose you invest $100 at 5 percent nominal annual interest. Compare the return under annual compounding and continuous compounding for one, five, and ten years. Repeat for 10 percent interest and for 20 percent interest (see Table A.1).

Note from the column of differences between continuous compounding and annual compounding that the difference is greater in proportion to the investment period and the difference is greater in proportion to the interest rate.

TABLE A.1 RESULTS OF ANNUAL AND CONTINUOUS COMPOUNDING OF $100 FOR VARIOUS RATES AND PERIODS

Rate (%)	Years	Compounded Annually ($)	Compounded Continuously ($)	Difference ($)
5	1	105.00	105.13	0.13
	5	127.63	128.40	0.77
	10	162.89	164.87	1.98
10	1	110.00	110.52	0.52
	5	161.05	164.87	3.82
	10	259.37	271.83	12.46
20	1	120.00	122.14	2.14
	5	248.83	271.83	23.00
	10	619.17	738.91	119.74

A.3 NOMINAL AND EFFECTIVE DISCOUNT RATES

To find the relationship between a nominal interest rate and the equivalent annual rate, equate the two future sums.

$$\begin{array}{l} \text{Future sum at } i \text{ interest rate,} \\ m \text{ periods per year, } n' \text{ years} \end{array} = \begin{array}{l} \text{Future sum at } r' \text{ interest rate,} \\ 1 \text{ period per year } n' \text{ years} \end{array} \qquad \text{(A.12)}$$

$$P(1 + i)^{mn'} = P(1 + r')^{n'}$$

Obviously, this relation is independent of the present sum, and because the comparison is on an annual basis, let $n' = 1$.

$$(1 + i)^m = 1 + r'$$

or

$$r' = (1 + i)^m - 1 \qquad \text{(A.13)}$$

Example A.4

Return now to Example A.2 and compute the annual effective discount rate at which interest is being earned.

$$m = 4 \quad \text{and} \quad i = r/m = 0.04 = 0.01$$

Thus (A.13) becomes

$$r' = (1 + 0.01)^4 - 1 = 0.0406$$

Thus a nominal annual discount rate of 4 percent, if compounding occurs quarterly, results in an effective annual rate of 4.06 percent.

The approach used in Eq. (A.12) can also be used to determine the effective annual rate obtained by continuous compounding.

$$\begin{array}{l} \text{Future sum at } r \text{ nominal annual rate,} \\ \text{compounded continuously} \end{array} = \begin{array}{l} \text{Future sum at } r' \text{ effective annual rate,} \\ 1 \text{ period per year} \end{array}$$

$$Pe^{rn'} = P(1 + r')^{n'}$$

$$e^r = 1 + r' \qquad \text{(A.14)}$$

$$r' = e^r - 1$$

Example A.5

Return to Example A.3 and compute the effective annual interest rate earned by continuously compounding the nominal annual rates of 5, 10, and 20 percent.

$$r' = e^{0.05} - 1 = 0.0513$$

$$r' = e^{0.10} - 1 = 0.1052$$

$$r' = e^{0.20} - 1 = 0.2214$$

It appears that for moderate amounts of money and moderate discount rates, continuous compounding does not have a great advantage over annual compounding, and it would of course have a smaller gain over quarterly or monthly compounding.

A.4 UNIFORM PAYMENT RELATIONS

There is no end to the variety of problems that can be devised using Eq. (A.1) or its variants. Indeed, one complaint about many introductory courses in applied economics is that they confuse things unnecessarily by introducing too many special cases and special formulas, resulting in a cookbook flavor. Note the following special case, however, which is worth treating because it is rather common.

Equation (A.1) deals with what is called the single-payment formula, that is, the value at some time in the future of a single sum P invested under a given set of conditions. Another common situation is that of equal amounts of money paid in or out at uniform time intervals, in effect, returned as an annuity. You could also imagine other situations such as uniformly increasing or decreasing payments, called uniform gradient problems, but this discussion is restricted to a series of uniform amounts. This arrangement is commonly used for repaying bank loans, mortgages, and other time payment plans.

Seek the value after n periods of an amount A invested at the end of the first period and the same amount invested each period for n periods. It may seem strange that the payments don't start at time zero, but the conventional approach is to pay nothing until the end of the first period.

Apply Eq. (A.1) repetitively to find the future sum.

$$F = A[(1 + i)^{n-1} + (1 + i)^{n-2} + \ldots + (1 + i) + 1] \tag{A.15}$$

The quantity in brackets is the geometric progression in $(1 + i)$, whose sum is well known. Thus

$$F = \frac{A[(1 + i)^n - 1]}{i} \tag{A.16}$$

In standard functional form Eq. (A.16) may be written as follows:

$$F = A(F/A, \ i\%, \ n) \tag{A.17}$$

Example A.6

"Neither a lender nor a borrower be." A friend borrowed $500 from you and agreed to pay you back at the end of the year with 10 percent interest. Now the friend has a new job and can afford to pay you a given sum at the end of each month. The friend can afford $50 monthly payments but feels that you should forgive the first month's payment. Should you stick with the original plan, or would you be better off with monthly payments?

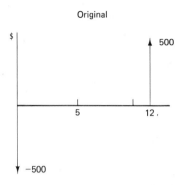

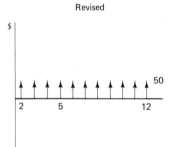

Figure A.2 Cash flow of lender under original plan and proposed revision.

Your two cash flows are shown in Fig. A.2. It is clear from Eq. (A.1) that you receive $550 at the end of the year under the original agreement. If $r = 0.1$ and $m = 12$, then $i = 0.1/12 = 0.0083$ and Eq. (A.12) yields

$$F = \frac{\$50 \left[(1.0083)^{12} - 1\right]}{0.0083} = \$628.16$$

But if the first payment is forgiven, you lose $50 plus 11 months' interest on that $50, or

$$\$50(1 + 0.1/12)^{11} = \$54.78$$

Thus the second plan yields

$$\$628.16 - \$54.78 = \$573.38$$

It appears that 11 monthly payments are superior by $23.38 to the single-payment plan, and you should accept the monthly payments.

What is the present value of a series of equal future annual payments? Suppose that an amount A will be paid at the end of each year for the next ten years. At a discount rate of 5 percent, find the present value of this stream of funds.

You can't use Eq. (A.16) for this problem because it tells the value of equal payments at the end of the investment period. You seek the present value, at time zero, of future payments. Of course, you could calculate the future value and reflect it back to the present using Eq. (A.1), but a more direct way exists. Eq. (A.1) may be rewritten to reflect the present value of a future amount by simple transposition.

$$P = F(1 + i)^{-n} = F(P/F, i\%, n) \tag{A.18}$$

A relation similar in spirit to Eq. (A.15) can be written for equal annual payments based on repetitive application of Eq. (A.18).

$$P = A(1 + i)^{-1} + A(1 + i)^{-2} + \ldots + A(1 + i)^{-n} \tag{A.19}$$

Economics books may have a relation for the sum of a series such as Eq. (A.19), but a cookbook isn't necessary. You can write a new relation by multiplying Eq. (A.19) through by $(1 + i)$ and subtracting the new relation from Eq. (A.19). When you do, most of the terms on the right side cancel, leaving

$$P - (1 + i)P = -A + A(1 + i)^{-n}$$

or

$$P = \frac{A[1 - (1 + i)^{-n}]}{i} \tag{A.20}$$

Equation (A.20) may be written in standard functional form as follows:

$$P = A(P/A, i\%, n) \tag{A.21}$$

Example A.7

What is the present value of a series of ten year-end payments of amount A assuming a nominal discount rate of 5 percent compounded annually?

In this example, $i = 0.05$ and $n = 10$. Using Eq. (A.21), the result is

$$P = A(P/A, 5\%, 10)$$

or,

$$P = \frac{A[1 - (1.05)^{-10}]}{0.05}$$

Thus the present value of ten payments spread out over ten years is not equal to $10A$, except at a zero discount rate. Can you show this fact using Eq. (A.20)? At a discount rate of 5 percent, its current worth is only $7.72A$.

EXERCISES

1. What nominal annual interest rate compounded quarterly will double an investment in five years?
2. A $5000 loan is to be repaid in six years with payments of $110 per month. What are the nominal and effective interest rates of the loan?

3. A loan shark charges interest at the rate of 1.5 percent per week. What are the nominal and effective annual interest rates?

4. In early 1981, *Newsweek* pointed to a U.S. House of Representatives member's statement that "decommissioning a single nuclear plant could cost as much as $100 million" [4]. The estimated operating life of a nuclear electric-generating station is 30 to 40 years. Some utilities have established sinking funds (money taken from earnings and set aside for future use) to cover decommissioning costs. How much money should a utility deposit annually in its nuclear plant decommissioning fund if it assumes a 40-year operating life, $100 million decommissioning cost, and a discount rate of 10 percent?

5. A pack-a-day smoker decides to quit on January 1, 1991, and at the end of each month put into a savings account the money he would normally spend for cigarettes ($30 per month). The bank pays him 6 percent per year, compounded monthly. How much will he have in the bank on January 1, 1992?

6. I want to buy a motorcycle. I have $500 available for a down payment and am willing to pay $100 per month for one year and $150 per month for the next year. The motorcycle shop will finance the unpaid balance at 1.5 percent per month. What is the most expensive motorcycle I can purchase?

7. The Equal Employment Opportunity Commission (EEOC) and Ford Motor Company settled a seven-year complaint. Ford must spend millions of dollars to make up for past discrimination against women and minorities [5]. Ford has to spend $383,000 each month over a five-year period beginning January 1, 1982. Assuming end-of-month payments, calculate the present worth (as of January 1, 1982) of this settlement if constant dollars are assumed (interest rate = 0 percent) and if the discount rate is 10 percent.

REFERENCES

1. D. G. Newman, *Engineering Economic Analysis,* rev. ed. (San Jose, Calif.: Engineering Press, 1977).

2. E. P. De Garmo, *Engineering Economy,* 4th ed. (New York: Macmillan, 1967).

3. L. E. Bussey, *The Economic Analysis of Industrial Projects* (Englewood Cliffs, N.J.: Prentice-Hall, Inc., 1978).

4. M. Sheib, "Reagan's Nuclear Reaction," *Newsweek,* January 12, 1981, pp. 62–64.

5. D. French, *Forbes,* February 16, 1981, p. 38.

The Times Mirror Calculation

There are four items in the cash flow shown in Fig. 7.1. Three are fairly clear, but the fourth is ambiguous. Thus we will calculate it two ways. The quotes are from the *Business Week* article cited in Chapter 7.

Item 1. "an upfront payment of $25M." *Upfront* means immediate, and the anniversary date is probably the date mentioned in the text, October 22, 1980. With no loss in generality we can assume for simplicity that the anniversary date is the end of the year (EOY) 1980.

Item 2. "semiannual interest payments at a 10 percent rate [on the $55 million] due in 1990."

Item 3. The $55 million due in 1990 (EOY assumed).

Item 4. "The remaining $15M interest free, payable in installments between 1991 and 2000" (EOY assumed). This quote is ambiguous. Normally annual payments are assumed, but in view of the interest payments in the period between 1980 and 1990, we could assume semiannual payments here.

Item 1

This item is obvious. Since the payment is upfront, its present value is $25 million.

$$P_1 = \$25 \text{ million}$$

Item 2

Each of the semiannual interest payments at 10 percent is on the total principal of $55 million. Note from Appendix A, Eq. (A.7), that at the moment of payment the value of each payment is

$$A_2 = (i/2) \times \$55 \text{ million} = (0.10/2) \times \$55 \text{ million} = \$2.75 \text{ million}$$

The annual interest is divided in half because the period is half a year. The payments begin six months into 1981 and continue through EOY 1990. Note that payments from items 2 and 3 both occur at EOY 1990.

Item 3

The $55 million principal payment is due at EOY 1990.

$$F_3 = \$55 \text{ million}$$

Item 4

This item seems open to at least two interpretations. Assuming ten annual payments first,

$$A_4 = \$1.5 \text{ million}$$

If semiannual payments are assumed,

$$A_4 = \$0.75 \text{ million}$$

Business Week is silent on the matter, but the semiannual option is more likely. Thus

$$P = P_1 + P_2 + P_3 + P_4$$

or

$$P = \$25\text{M} + \$2.75\text{M}(P/A, \, i\%/2, \, 20)$$
$$+ \$55\text{M}(P/F, \, i\%, \, 10) \qquad \qquad (\text{B.1})$$
$$+ \$0.75\text{M} \, (P/A, \, i\%/2, \, 20) \, (P/F, \, i\%, \, 10)$$

The final term in parentheses in Eq. (B.1) is necessary to reflect the payment back from EOY 1990 to time zero if standard tables are used. Thus if $i = 10$ percent,

$$P = \$25\text{M} + \$2.75\text{M}(12.462) + \$55\text{M}(0.3855)$$
$$+ \$0.75(12.462)(0.3855) = \$84.08\text{M}$$

What difference does it make if the payments in the 1990–2000 period are annual or semiannual? The first three items in Eq. (B.1) remain the same, while item 4 becomes

$$P_4 = \$1.5\text{M}(P/A, \, i, \, 10)(P/F, \, i, \, 10)$$
$$= \$1.5\text{M}(6.145)(0.3855) = \$3.55\text{M}$$

and the value of (B.1) becomes \$84.03 million. However, a small difference between two large numbers is subject to large error. If we calculate the PW of the extra interest directly and bring these ten payments back to time zero, we see that the present value is about \$88,834 at 10 percent.

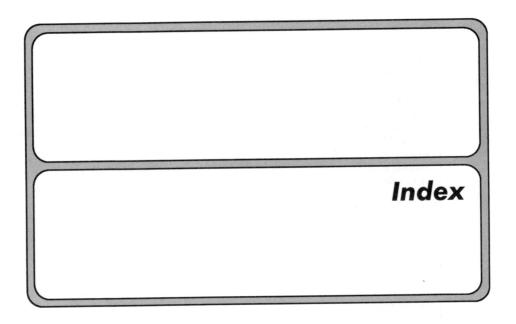

Index